Author's note

Thank you to my wife Christie for her never-ending help, patience and constructive criticism.

Thank you to old friends, some of whom are mentioned in this book: Duncan, Diane, Winston, Bunky, Gordy, Nookie, Carron, and any others I may have forgotten, and Zach Sagurs of the Bermuda Classic Bike Club for the photos of old Bermuda mopeds. It has been 60 years since these events so I may have forgotten some, and those that I have mentioned may be slightly warped by the passage of time, but I have tried to be as accurate and honest as possible.

My comedic doodlings are meant to add a bit of levity to the narrative and should not be judged for any artistic talent.

Most of all, thank you to Noel for being my friend in these adventures. You left us far too early, my friend. Who knows what you might have achieved?

>>0<<

All rights reserved.
No part of this publication may be reproduced or copied without permission of the author or the publisher.
2020

Born in the UIK while a nasty piece of work called Adolf was trying to wipe us off the face of the earth, I lived near the south coast as a toddler with bombs falling all around the neighbourhood. I can actually remember smoking 'shrapnel' on the doorstep!
My Scottish father served in the Royal Navy and after he was torpedoed my mother took me to her parents' house in Virginia,USA for a few years and then got a job in Bermuda after the war ended.
Ultimately I ended up working for the Bermuda Zoological Society at the Bermuda Aquarium,Museum & Zoo before retiring to Scotland. While at BAMZ I was inspired to write my first book, Greenhouse 2065-The Wave. Having done that, I thought it might be fun to give people an insight into my young adult years in Bermuda.
I hope you enjoy it.

John McGill
Dumfries,Scotland
2020

Adventures with Noel
by
John McGill

Contents

BERMUDA MAP 4

THE BEGINNING : SINGING ANGELS 5

MOPED 1- VeloSolex 9

MOPED 2 – Mobylette 16

THE FISHING TRIP 18

THE HYDROPLANE 29

THE EXPLODING OCTOPUS 36

MOPED 3 - Zundapp 44

THE PARTY 45

MOPED 4 – Motom 52

THE WEDDING 56

JAWS 62

MOPED 5 – CYRUS 73

WHITE'S ISLAND 76

THE RACE 80

THE MOON 88

BIKE TESTING 91

WANT TO GO HITCH-HIKING? 97

THE PLAN	102
THE FLIGHT	113
MADRID-The River Bank Hotel	123
MADRID- The Angry Gardener	129
MADRID- La Posada	136
GOODBYE MADRID	145
GUADALAJARA	152
ZARAGOZA TO MILAN	161
MILAN	174
COMO	180
THE ALPS	186
LYON	196
THE CAFE	198
PARIS	206
CALAIS-DOVER	207
ENGLAND	213
BLACKBURN	215
EUROPE	228
BERMUDA	234
THE END	236

BERMUDA MAP

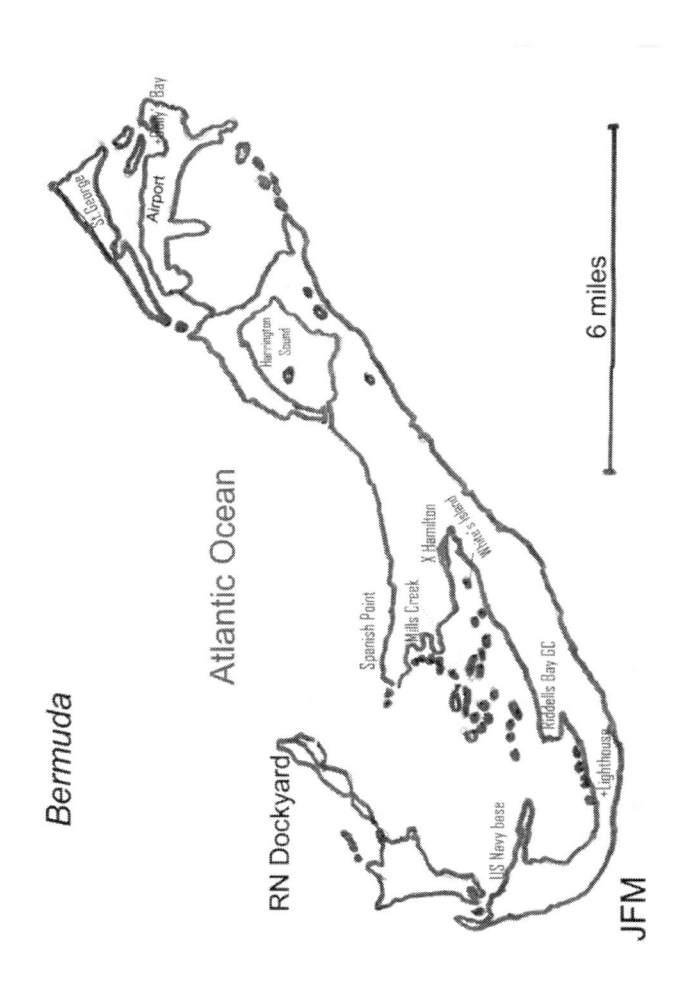

THE BEGINNING : SINGING ANGELS

What inspires a person to sit down and write a book about things that happened 60 years ago? I have no idea what made me think of doing it. Perhaps it was because I have just finished my other two books and, discovering how simple it is to self-publish, I was thinking "Now, what else can I write about?" I could write about my time in Africa, driving by myself (idiot) from Nairobi to Amboseli through swirling dust tornadoes and past hitch-hiking Masai tribesmen carrying their assegai spears until I could finally gaze at Mount Kilimanjaro through its swirling clouds, or I could write about driving along Route 66 in America or cruising the Norwegian fjords.

I can't say it came as some sort of lightning-flash idea, but it did suddenly dawn on me that I could write about my time during my youth with my best pal Noel and the fun, crazy and sometimes amazingly stupid things we did together. Noel and I met in school, Saltus Grammar School, said to be the best in Bermuda at the time I attended, although I am sure its lofty reputation had nothing to do with the fact that I was a pupil there, nor would its reputation have been enhanced by Noel's presence! We both always seemed to be getting into trouble.

I think we became friends because we had a lot in common. We were both English, although I had arrived in 1948 from the UK via Virginia with my mother after my father had been torpedoed in World War 2, whereas Noel had arrived much later after his father had been appointed to a senior position in The Bermuda Telephone Company, known to locals as Telco. This made us two 'limeys' in a school full of 'onions', a nickname earned by Bermudians from the days in the 1800s when the island exported tons of the sweet Bermuda Onions to America. To this day Bermudians are still proud to be called 'Onions'.

In many ways, Noel and I were also different. I was 6 feet tall and skinny. He was slightly shorter than me and, although I would not want to call him plump, he definitely carried more weight than I did. With his short brown curly hair and blue eyes, coupled with a deceptive angelic mischievous grin, you could believe the claim that he had sung at Queen Elizabeth's coronation, a fact that was often announced by his proud mother. There was a total of several hundred boys drawn from church choirs all over the UK to sing at the coronation and they had rehearsed separately to the same music before they were eventually all brought together at the wonderful pageant in 1953. The boys would have been seated all around the choir section of Westminster Abbey and expected to behave themselves for the duration of the long service. The service itself lasted for three hours, and the boys had to be there long before that so that they could be properly dressed and prepared. They were given a 'survival kit' of sweets, chocolate, bread and butter, as well as a small bottle of milk and they were advised to hold onto the empty bottle once they drank the milk "In case you need it later." There was even a rumour that one choir boy had thrown small paper darts at the other boys during the coronation service. Noel never owned up to being the perpetrator, but I definitely would have believed it if I had been told it was him.

As for myself, I could not claim that I sang at the Queen's coronation, but I did sing at a few weddings while in the choir at St. Anne's church in Southampton Parish, Bermuda! I didn't throw darts at my fellow choir boys, but we did fall about laughing at silly jokes, such as the one I always remember about the priest who trained his parrot to strike matches and throw them from the rafters of the church just as the priest, in the middle of his sermon, bellowed "And fire came down from heaven!". No matches came down from heaven, and the priest bellowed out again, "And fire came down from heaven!" and from the shadows of the rafters came the reply "Squawk! The dog peed on the matches!"

So there we were, thrown together and becoming friends on a small jewel of an island in the Atlantic. What times we had!

MOPED 1- VeloSolex

Every teenager in Bermuda looked forward to the day he or she turned sixteen. That is the age at which you can legally ride a 'moped' in Bermuda. Although not well-known in North America, mopeds were widely used in European countries and were ideally suited to Bermuda's narrow roads and 20 mph speed limit. They were exactly as the name suggests: motor-assisted pedal bikes. Bermuda is a very small island with narrow and twisting roads. Regulations have always stated, wisely, that there can only be one car per household. This reduces traffic congestion on the island's roads. Tourists could not hire cars, only mopeds. Many a fortune was made by enterprising Bermudian businesses renting mopeds to tourists. These days, tourists do not flock to the island in the same numbers that they did when I was a teenager so there are less of them on the roads, although quite a few come by cruise ships now, but they don't tend to hire mopeds. It is still a common sight to see Bermuda businessmen riding their moped or scooter to work dressed in Bermuda shorts, knee-length socks, tie and a blazer, while the family car is used for school runs and shopping..

Prior to coming of age for riding mopeds in Bermuda, I had a 3-speed Raleigh pedal bike which I rode to and from school every weekday. I took pride in the fact that on several occasions when the wind was right I could pedal the seven miles from my house in Southampton to Saltus Grammar School located behind the city of Hamilton in 20 minutes! I only achieved this by riding along Harbour Road in the mornings with a good westerly wind at my back. I don't think it would be possible to do it in these modern times because the density of traffic on Bermuda's roads is so much greater than it was in the 1950s.

I had every make of moped that was available at the time, which was 1956 onward. I was a few months older than Noel and therefore I was 'on the road' sooner than him. It did not make a lot of difference because, by the time I was buying my first bike in our mid-teens, his parents had decided to send him away to school in Canada, which meant that we only saw each other during Easter, summer and Christmas holidays.

My first bike was an old VeloSolex and was the cause of my first run-in with the law. A lady friend of my mother was getting rid of her Velo and offered it to her at a very low price. As soon as I heard about the bike I pestered my mother until she bought it. My mother was reluctant to buy it with good reason. I was not 16 yet. She bought it, though, and naturally, I could not resist taking it for a test ride.

I should explain that the VeloSolex, which I so desperately wanted, was a joke among my peers. It was a pedal bike with a heavy black frame and no cross-bar. It had a tiny 49cc engine stuck on the front that, quite literally, rested on top of the front tyre. The engine could be pulled up slightly by a handle so that it disengaged from the front tyre and it then became a rather ungainly pedal bike. When you wanted to start the engine you just pedalled to get going and then dropped the engine down onto the wheel by pushing the lever forward. This made a serrated roller rest firmly on the tyre and started the engine, which then started to turn the front wheel. It had very little power and only did about 15 mph. It was not exactly The Bike of Champions. What was worse was that on wet roads the wheel would get slippery and the engine would slip and you had to pedal furiously, even on the flat bits of road, especially after the tyre had been worn down until it was almost completely smooth.

 Bermuda is a small island with a subsequently small population and you could not go very far without bumping into somebody who knew you or your family. This meant that the Parish Constable knew everybody on his patch. On my test day, at the grand old age of fifteen years plus ten and a half months, I ventured forth, got my Velo going and raced along at 15 mph with the wind in my hair. I was flying!

I had only gone about two miles from the house, pedalling furiously up the hills (Bermuda is very hilly) when I heard a deep-throated rumble behind me. My heart sank as I looked around and saw our Parish Constable, 'Squid' Moniz on his police bike.

It is a remarkable fact that if you look in a Bermuda Telephone directory you will see many entries with the listed person's nickname in brackets. I have seen an entry listed as Curtis, Charles (Centipede) and our parish constable that was now pulling alongside me was known as 'Squid' Moniz. I would never think of addressing him by his nickname unless I wanted to incur his wrath, which I did not want at this moment.

Mr Moniz pulled me over to the side of the road and, using the age-old police tactic of the 'soft approach' said, "Hello John, out for a little ride, are you?"

Oh no! He knew me! "Er, yes sir, Mr Moniz sir," I muttered.

My nervously polite response must have satisfied him because, in a kindly way he uttered these terrible words, "Now John, I know your mother," at this point, I knew that I was in trouble. Everybody knew my mother, she was a stalwart of the local church, Sunday School teacher and choir member.

Mr. 'Squid' Moniz continued, "Now John, I know your mother and I happen to know that you're not 16 for another month, so you shouldn't be riding that Velo, should you?"

"N-n-n-no, sir. Sorry, sir." I stuttered.

I realise that in these modern times a fifteen-year-old boy who is almost sixteen would probably have laughed at the policeman and shouted some expletives at him, thrown the Velo at him and pushed through the bushes and made his escape. However, I had been brought up to respect my elders and authority figures.

"Well now," said Squid the beneficent parish constable, "I suggest you turn your bike around and ride back home and do not ride it on the public road again until you are sixteen, is that clear?"

Overwhelmed with relief and gratitude, I resisted flinging my arms around him and instead I turned my bike around and headed for the comforts of home.

To this day, I have never forgotten the surprise that this man knew who I was and how old I was. Those days are gone now, I believe, but at the time, when I thought about it, it was reassuring. For the next week, I waited for my mother to summon me and ask for an explanation as to why I was riding my Velo illegally. To my surprise, she never did confront me, and I realised that my new friend Squid had never told on me...and I never did find out why he was called Squid. It probably had something to do with fishing.

The "Velo"

MOPED 2 – Mobylette

I was so embarrassed by my VeloSolex that it only lasted a month before I convinced my mother that it was broken and I needed a new bike. Before I became sixteen I had been working on Saturdays at a shop called Archie Brown & Son. I helped moving stock around the shop and every morning I swept the sidewalk outside the Front Street shop which was fun because I got to say 'Good morning' to all sorts of people who were walking past and all the people that I worked with at the shop were very nice. Especially nice was a rather voluptuous French lady who worked in the perfume department. She was dressed quite often in a very low-cut dress and I remember all the young male staff seemed to find it necessary to walk in her vicinity in the hopes of peering down the front of her dress!

I was allowed to keep all my earnings and after a month or so I moved on to my second moped. This time I really moved up in the world by purchasing a very old and very used Mobylette. This French machine was a REAL moped with a small 49cc motor built onto the frame with a chain drive to the rear wheel just like a big motorbike. The only embarrassing thing about this bike was that a lady had owned it and it had a wicker basket attached to the handlebars at the front of the bike where I imagined a handbag would have been carried (or ET!). It was so obviously an old lady's bike that I couldn't wait to get rid of the wicker basket and replace it with some cool saddlebags.

By now, Noel's parents had decided that he should go to school in Canada but we still got together on the holidays when he came home. In the meantime I rode my Mobylette to school, feeling very thankful that I no longer had to pedal the seven miles to and from school every day, which is what I had been doing for the last few years. The Mobylette didn't go very fast but it didn't matter, with the Bermuda speed limit being 20 mph, and I only had to pedal up the steepest hills. It lasted until my seventeenth birthday, at which time I left school and applied for a job at American International Company and felt very pleased with myself when they hired me on the spot. I was very good at figures and this is what impressed them, I think. It was 1957 and I was earning the huge sum of £8 a week! For a while, I was happy with my Mobylette. It had a real engine on the frame, a gas tank under the seat and built into the frame. The Bermuda mopeds all took a special fuel mixture of petrol and oil and they got very good mileage. For now, I was happy to ride it to work every day and it cost almost nothing to run.

The "Moby"

THE FISHING TRIP

That year Noel got a job at a boat rental business for the summer that operated out of a location near his house. The boats were very basic 16-foot runabouts with a small outboard engine that tourists could hire for the day to explore around the islands on the understanding that they did not venture outside of the area just to the west of Hamilton Harbour.

It was inevitable that a person like Noel would see the possibilities that presented themselves when he was put in charge of the boat rentals.
There were about six boats and he soon knew which ones were reliable and which ones were troublesome. He only rented out the troublesome ones after the others had already been rented. This cut down on his overtime hours spent in repairing them. It also meant that he knew which one he wanted to borrow for a little night-fishing. The owner of the boats had no problem with Noel borrowing a boat, but he had no way of knowing that Noel had invited a few of us to join him.

Noel was an easy-going and fun guy to be around and he soon attracted a mixed bunch of friends. Apart from myself, there was Douglas, a cheeky chap who went on to become a deep-sea diver; Ronnie, a cheerfully nice guy who became a marine contractor; Gordy with a great sense of humour; Lenny whose father was in the hotel business and later returned to the United States and One-Shot (I don't know) with a wicked sense of humour and an ability to get into mischief; Roger who was a keen sailor and became a government health inspector and Nookie (don't ask)a great guy who later set up his own business and became a fine golfer. Others came and went, but these were the main group.

One night a bunch of us went to Noel's house which had a dock on Fairyland Creek, a branch of Mills Creek where the rental business was located, and at about 6 pm he nosed into the dock with the borrowed boat and we all piled in with some fishing gear and we set off for our night-fishing adventure.
It was high tide as we cruised slowly out of Mills Creek and set course for the Little Sound. Bermuda is shaped like a fish hook and the Little Sound and Great Sound were in the bowl formed at the bottom of the hook with the peninsula of the U.S. Navy base separating the two sounds.

The US Navy had seaplanes based there and they used to land and take off from a stretch of water that was marked by floating markers. The aircraft usually came in over the hill where the Gibbs Hill lighthouse was located, using the lighthouse as a guide, and landed in the area where we were headed.

We weren't worried about the planes because we were going to stay clear of their landing area if they happened to be doing night-landings. Well, that was the plan.

Leaving the shelter of Mills Creek we headed past the various islands at the entrance to Hamilton Harbour. On our right, in the distance to the west, we could see the lights of a large cruise ship at anchor in Grassy Bay. There was lots of chat and somebody had brought along a bottle of cheap wine which got passed around as the banter bounced back and forth with great merriment. We all enjoyed telling jokes and everybody was in a good mood. The weather was calm and the sea was smooth as we left the islands behind us and entered the Little Sound. To our left, we could see the lights coming on in the houses along the shore at the edge of the Riddells Bay golf course. To the right of us, the lights of the floats marking the landing area of the seaplanes came on. It was becoming twilight as we neared our destination, a big seaplane mooring buoy. There were about three of these big buoys in the sound that could be used by the US Navy when they needed to anchor seaplanes in the sound temporarily. During the war, Morgan's point, which was the local name for the US Navy base, had been very busy with dozens of seaplanes there, pulled up on the huge ramps they had built or moored on the big buoys, but now it was pretty quiet with the odd PBY Catalina patrol planes flying in and out, but mostly there were one or two P5M Marlins flying patrols and doing practice take-offs and landings in the sound.

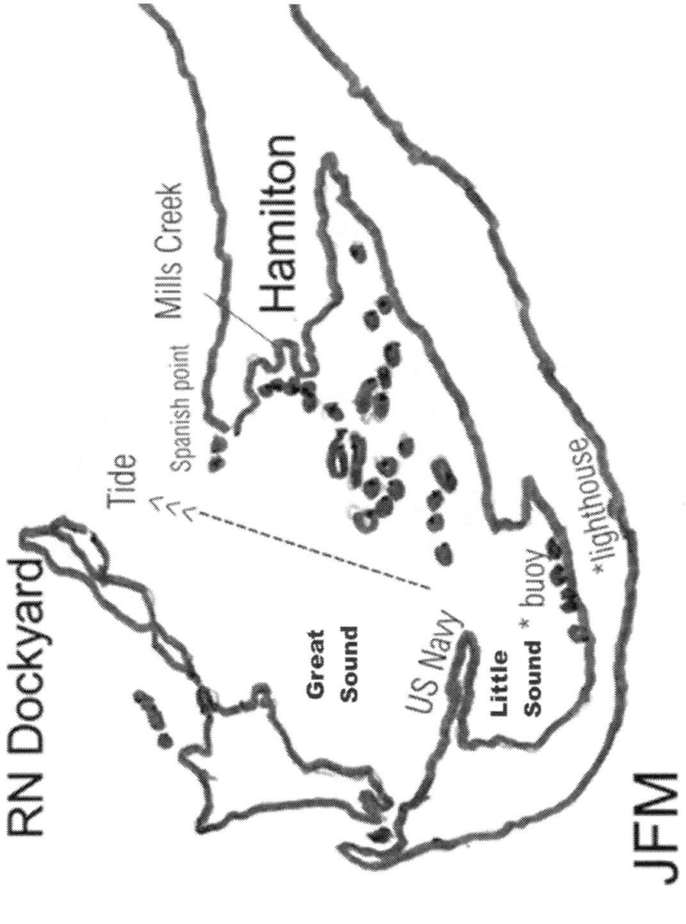
RN Dockyard
Tide
Spanish point
Mills Creek
Hamilton
Great Sound
US Navy
Little Sound
* buoy
*lighthouse
JFM

The buoys were quite big, about six feet across, round metal things that floated almost four feet high in the water and moored by chains to large weights at the bottom of the sea. They had a large ring in the centre on top so that aircraft could be secured to them. What made them so attractive to us was that they had been there for years and had lots of weeds and barnacles on the bottom and they attracted lots of fish that congregated beneath them.

We approached the first buoy and one of the guys managed to reach up and tied us to the ring on the top. Noel, as the captain, was at the stern of the boat in charge of the outboard motor. Once we were securely tied to the buoy he turned the engine off.

It was calm and peaceful. If you've never been on the water at night with no engine noise, just the water lapping at the side of the boat, it is hard to understand how nice it is. We all relaxed, a few of us threw some lines over the side in a pretence of fishing but we were really there to have a good time, chatting among ourselves. Once in a while, somebody caught a fish, but it was usually a grunt or a snapper, not big enough to take home, so it was lucky enough to be thrown back into the water to swim free and grow bigger.

All around it was dark, but not pitch black. We could see the land about half a mile to the south, rising steeply to where the lighthouse, now lit, shone its warning beacon far out to sea. It could be seen by ships as far away as 26 miles on a clear night. I told the guys about my step-uncle being a lighthouse keeper when I was a lad of twelve, and how I used to walk up to the lighthouse when he was on duty and beg him to let me wind up the weight. That was back in the days before it was connected to electricity when the light came from a pressurised gauze lantern and the power to turn the wheel was provided by a huge weight that hung down the centre of the lighthouse through a huge central pipe. The weight had to be wound up every night before the light was turned on, sort of like a grandfather clock. The handle or crank that was used to winch the weight up was almost as big as me and it always wore me out long before I was able to get it wound all the way up to the top. As I told the story we laughed because I realised, now that I was older, that he must have thought it was a great joke for me to beg him to let me wind it up, and him sitting there watching me struggle with the huge handle.

As it got later we decided to call it a day and head home. Just then we heard a sound and, looking up, saw lights appear over the hill near the lighthouse as a seaplane dropped down like a huge dark bird and passed over us to land in the sound only a hundred yards away. It was time to leave!

I was sure I heard Noel say, "Doug, untie the boat once I start the engine."

Unfortunately, Douglas later swore that all he heard was "Doug, untie the boat," ...so he did. Noel was bent over the engine preparing to pull on the start rope and did not notice. He pulled on the start rope and the engine stuttered and then remained silent on the stern, quiet and unresponsive.

Noel muttered something and gave another pull. Nothing happened. We all looked at each other in the dark, nobody seemed to be smiling. At this point, it became obvious to everybody that we had drifted away from the buoy because Doug had untied us. There was very little wind but we were noticeably drifting away from the shore in the general direction of the dockyard and after that, the open ocean. Noel looked around and said, "Shit Doug, you weren't supposed to untie us until I got it started!"

"No way! I heard you say 'untie the boat'!" this pointless conversation went on for a few minutes and then somebody else got up and tried to start the engine by pulling on the start rope. The engine had decided it was not going to start and after we had all had a go at pulling on the rope we realised that we were "up the creek without a paddle", literally. We had no oars, no safety equipment. We didn't even have an anchor. By now we were out in the middle of the sound and had absolutely no emergency equipment. This was in the days when we didn't have mobile phones or satellite navigation. We were on our own and nobody knew where we were except us, and we weren't too sure either. What we did know was that the situation did not look good!

A half an hour had passed when we heard a familiar sound. Looking back in the direction of the lighthouse, we saw the lights of the seaplane appear again out of the dark sky and start to descend toward the water for his second practice landing. He was coming right toward us. We all started to

lean over the side and paddle frantically with our hands. The boat barely moved. Noel reached down and started tugging at one of the floor-boards. Rip! Up it came, a four- foot long plank. He leaned over the side and started to paddle furiously. The boat began to move and the plane descended lower, coming right for us. Somebody else ripped up another floorboard, then another, and another.
Soon there were four of us paddling like crazed demons. The floor-boards were turning the water around the boat to a foaming white and we flew along at a magnificent 5 mph. Unfortunately, the seaplane had a landing speed of approximately 100 mph. We never had a chance.

Suddenly everything was quiet, nobody was paddling. Every eye was fixed on the landing lights of the aircraft. It was close enough to see. It came lower and lower. Whenever we used to watch the US Navy seaplanes taking off and landing it was from a safe distance on the shore and they didn't look so big, but now the fact that they were a hundred feet long, had wings one hundred and twenty feet wide and weighed about twenty tons became extremely obvious. Nobody said a word. I still remember the moment, to this day, that it swept over us no more than ten feet above our heads and touched down on the water twenty yards away from us in a huge cloud of spray that drifted back over us.

It dawned on us all at that moment that the tide was still going out and it was taking us right into the middle of the landing zone. Nobody said it, but I think we all decided at the same moment that the thought of several tons of metal seaplane landing on top of us was not a pleasant one. The normal tide difference between high and low tide in Bermuda is only about four feet, but the volume of water flowing out of the Little Sound was still enough to carry us along quite swiftly. We all started to paddle fiercely toward shore as hard as we could, and we did manage to get clear of the landing zone but we were still being swept toward Spanish Point, the last bit of land before we were carried out to sea.

Suddenly somebody cried, "Look, the cruise ship tender!" and we all looked up from our frantic paddling.
Sure enough, coming toward us was one of the large government ferries that were used to shuttle cruise ship passengers back and forth to Hamilton. Some cruise ships tied up alongside Front Street in Hamilton, but the larger ones had to anchor out in the Great Sound when the Hamilton docks were too busy.

It was just after 11:00 pm and this would be the tender returning to Hamilton after taking the final batch of passengers to the ship for the night.
We were saved!
As it approached we started to jump up and down and yell and wave our hands like demented puppets. Slowly it dawned on us that it would pass approximately fifty yards behind us as it headed for Two Rock Passage into Hamilton Harbour. Some of the group had matches for lighting the occasional cigarette and we now began striking the matches and throwing them into the air. How pathetic they looked. They flared brightly for two seconds, flew up for about three feet, and then fell into the ocean.

Suddenly a searchlight was turned on aboard the tender and it swept the sea around us and then fixed on us. We all jumped up and down and waved. The tender gave us three cheerful blasts on its ship's horn and then the light was switched off and it carried on into Hamilton Harbour.
We couldn't believe it. They must have thought that we were in no distress and had simply been waving at them! We all slumped down in our little boat, floorboards scattered around and a useless engine sitting silently on the stern.
"Come on, keep paddling," said Noel and we all took turns paddling. It was midnight and the outgoing tide seemed to be swirling around and pushing us closer to Spanish Point. At 2:30 am an exhausted group of shipwrecked mariners

dragged themselves ashore on Spanish Point and sat exhausted and silent.

"Well, at least we made it," somebody said. Nobody bothered to answer. We were all too tired. It was a good hour's walk back to Noel's house where we all had our bikes.

By the time I got home it was 4:00 am, and I still had to go to work!

I managed to get up and go to work, but my supervisor Dave Vaughan found me asleep at my desk, slumped over a spreadsheet at 11:00 and after asking me what had happened, he took pity on me and sent me home.

THE HYDROPLANE

As far as I can remember, Noel didn't have very much luck with boats, even though he was a pretty clever guy when it came to most things.

I remember, sometime after the fishing trip, he came to me and said, "I need a ride down to St.David's."

"OK," I said. "When, where and why?"

"There's a guy down there who has a very fast hydroplane for sale and I want to buy it," he said.

A funny thing about Bermuda is that even though St.George and St.David's are north of Hamilton, locals say they are going 'down' to St.George, and if going south to Somerset, they say they are going 'up' to Somerset. This dates back to the days when most travelling was done between the two ends of the island by sailboat, and the prevailing wind was almost always from the southwest, which meant that you were sailing downwind to St.George and upwind to Somerset. When Noel said he wanted to go 'down' to St.David's, it made complete sense to me.

What didn't make sense was why he would want to buy a fast hydroplane.

"That guy Jimmy at the marina was telling me he's got the fastest boat in Bermuda and I bet him £20 he doesn't."

"Why would you bet him when you don't even own a boat?" I asked.

"Because I had heard about this fast hydroplane in St.David's that is up for sale," he replied.

That's why, on my Thursday afternoon off, we headed to St.David's on my bike, Noel riding behind me, to buy a boat we had never seen so he could win his £20 bet.

We arrived in St.David's and after getting lost a few times we finally found the dock where the boat was awaiting inspection. The dock was down a little path between some bushes and it was simply a wooden jetty on somebody's property on Dolly's Bay. I had been expecting a proper cement dock, more like the public ferry docks scattered around the island.

We parked my bike near some bikes that were already there, and looked down at the waterside where a motley group of fellows were gathered, looking down at a rather brightly-coloured little boat. They were a very rough-looking bunch of guys. At their centre was a wild-looking short chap with what looked like a Mohawk Indian haircut, with just a wide row of hair along the centre of his head standing straight up for about two inches.

"What's this guy's name?" I asked Noel.

"Crazy Horse," he replied cheerfully.

I looked at him, then looked at the gang of guys who were now looking at us as we approached. Wild hunting dogs looking at a couple of juicy gazelles came to my mind. Noel acted like he was completely at home with the situation. Mr Mohawk, a very lean and tough-looking little guy came forward. "You No-ell?" he asked, smirking as he purposely mispronounced the name.

"Yeah man, I'm Noel." He replied and shook the guy's hand in some sort of ritual fist-bumping. It never ceased to amaze me how easily Noel mixed with everybody he met. "I guess you must be the man, Crazy Horse, yeh?"

The little guy puffed out his chest. One of his pals snickered and said, "Yeh, he be one crazy-assed bastard."

All his other friends laughed at this amazingly funny joke. We all gathered at the jetty's edge and looked down at 'the fastest boat in Bermuda' floating happily below us.

It seemed to be painted in every colour of paint that could be bought at the local hardware store. It looked like a floating rainbow.

Hydroplanes were, at that time, traditionally about eight to nine feet long, made of plywood and very light-weight. They had a flat bottom with a mini hull on each side that formed a 'tunnel' along the length of the boat. The concept was such that with a light boat and small engine you could obtain surprisingly high speeds because the air got under the boat, flowed along the tunnel and lifted it onto the twin hulls. There was only enough room in the narrow cockpit for the pilot and he would lean forward to keep the bow down so the boat wouldn't flip backwards as the wind started to get underneath the boat as speed increased. They raced them in America, we knew, and we had heard of them flipping upside down at speed.

This boat that we were looking at was obviously homemade, probably by a man known as Crazy Horse, which fact did not reassure me. The original idea was to have a very light boat and a small engine. This boat looked very light, but instead of the traditional 25 hp outboard, Crazy Horse had put a 50 hp engine on the stern. T

This could be very interesting, I thought.

The bow of the boat was sticking up in the air, the stern forced down by the weight of the over-sized engine. I couldn't help notice that one of Crazy Horse's pals had his foot resting on the bow, probably to stop it from lifting any higher. I looked at Noel. He looked at me and smiled. He turned to Crazy Horse.

"Do I get a free trial run before I buy it?"

"Yeah man, go ahead, just ease it away from the dock, OK?" replied the little guy. He seemed a bit nervous.

Noel was a pretty solid guy, and as he stepped into the boat, the bow tried to push upward and the fellow with his foot on the bow pushed down harder. It all seemed like a well-rehearsed act.

The engine was obviously in good condition because it started with only the slightest tug on the start rope. The bow rope was untied from the jetty and thrown into the boat and the guy on the bow pushed it out from the dock.

The engine was idling in neutral with the bow pointed away from the dock. Noel told me later that he knew it was going to be a disaster as soon as he saw the boat.

He pulled the gear lever forward and gave a little twist to the throttle grip. It was only a small amount of twist but the stern dug in and settled lower and the bow rose almost straight up, pointing at the sky. The boat went nowhere. The Atlantic Ocean, well, most of Dolly's Bay, started to pour over the stern and the boat slowly sank stern first. The engine kept going for ten seconds and then gave up. The only thing that kept the boat from sinking right to the bottom was the air trapped in the bow. The hydroplane rested with the prop on the shallow bottom.

At this point, I decided that a hasty retreat might be advisable and I started to ease slowly back up the path toward where my bike was parked. Everybody was standing with their mouths open looking down at the rainbow boat. Fortunately, the water was not very deep and Noel was able to stand when he jumped out and threw the bow rope to one of the guys on the jetty. Wading ashore, he quickly walked past Crazy Horse and caught up with me, and we strode up the path toward my bike. He looked back at Crazy Horse who was yelling at his pals, waving his arms in the air and ordering them to jump in and save his boat.

"If you think I'm buying that death trap, you really are crazy," Noel said. We increased our pace and quickly jumped on the bike and headed out of St.David's, leaving behind us the outraged cries of an extremely deranged Crazy Horse hurling some very colourful words in our direction.

A week later Noel told me that he didn't have to pay Jimmy the £20 because another guy with a faster boat had beaten Crazy Horse in a race. Our trip to St. David's had been unnecessary but it had been very interesting indeed.

JFM

THE EXPLODING OCTOPUS

 We had all gathered together down on the dock at Noel's house. His parents were away. I can't remember where they had gone or for how long, but an empty house meant we could gather and have fun without disturbing anybody. It was a great setup, with a large dock on Fairyland Creek. We could swim right from the dock and there was a very large shed with ample room for a party as well as workbenches and room to work on our bikes.

Naturally, with no supervision, there was plenty of beer and lots of laughter. I think it was Doug who found the octopus. It would be him because he went on to be a deep-sea diver. Anyhow, he came to the surface after splashing around at the front of the dock and said, "Hey you guys, I found an octopus."

Sure enough, he held his hand up and wrapped around it was an octopus. It was pretty big too, not a tiny one. He climbed up onto the grassy area by the dock and we all gathered around to inspect the creature.

I want to state right here and now, that in our later years as enlightened people , I and the rest of my friends would have returned the poor animal back to the water, but I can only blame the party atmosphere and the beer for what happened next.

We were so curious and eager to examine the octopus that the poor thing died. I had never seen one before in the wild and I don't think any of the others had either. It was such a surprise to find it living right there under the dock where we were gathered.

Once we realised it had died, we were feeling very sad and a bit guilty, but then somebody suggested that the Portuguese ate octopus all the time. We stood in a circle around the octopus lying on the grass in front of the shed. With its tentacles splayed out, it was the size of a large dinner plate and a grey-white with brown spots. There was some ink coming out of it also.

"How do you cook it?"

"I think you can just boil it."

"No, man. You have to chop it up and fry it up with spice 'n stuff."

"Yeah, like onions and other stuff." We were not expert cooks.

"I heard they are tough and you have to tenderise them."

"Whaddaya mean, tenderise?"

"You know, like with steak, they hammer it until it's tender."

"Oh, yeah, right. Get a hammer from the shed."

"No way, not that kind of hammer, they use special hammers with a big flat head."

"OK. We could use one of those blocks of wood under the shed."

"Well, we can't just hammer it on the grass. It'll just bash it into the dirt."

"OK. Put it on a piece of wood on the concrete of the dock."

If you had come down to the dock that day you would have seen five guys attacking something on the dock with a lump of wood. Octopuses are amazingly resilient and even after half an hour of hammering it was still looking quite healthy, apart from being dead, that is.

Eventually, we all paraded up to the house with Noel carrying the bashed octopus on a piece of wood and the rest of us in line behind him so that we looked like a funeral procession. Once in Noel's mother's kitchen, the discussion continued.

"It's gonna take a big pot."

"We should add onions and salt and pepper."

Noel rummaged around the kitchen looking for the necessary equipment and ingredients. It took a while because he had never actually done any cooking in his own house, or anywhere else. Nor had any of us, for that matter.

Finally, we had a huge pot, half-filled with water with salt and pepper, with some cut-up onions bobbing around in it. Noel lifted the octopus and lowered it into the pot, but the tentacles kept hanging over the side. Using one of his mother's big wooden spoons, he finally pushed each one down into the pot and firmly pressed the lid down on top. We lit the gas under the pot and somebody said, "Now what?"

"Now we wait," said Noel. "Anybody want a beer?"

We all trooped into the living room and sat around while he put some cool music on the record player. It was a very relaxed atmosphere, the beer was cold and all was right with the world. After half an hour, somebody said, "I wonder how the octopus is doing? Maybe we should look at it."

As if in answer, there was a very loud bang and then a clattering sound from the kitchen. We all jumped up and looked at each other and then there was a mad rush toward the kitchen.

One-Shot, at the front of the rush suddenly stopped and said "Oh shit!"

Noel pushed past him and said "Ohhhhhh man!"

All of us had pushed through the door by now and looked at the scene in front of us in amazement.

On top of the stove, the pot bubbled away quite happily. The octopus seemed to be peering over the edge, all eight of its tentacles were hanging over the side of the pot, dripping a dark brown liquid onto the stove-top. The lid was lying on the floor in the middle of the kitchen, still rocking. There were dark splotches everywhere, on the floor, on the kitchen countertops, and all over the ceiling. There were even dark brown inky splotches on the lovely white curtains at the window.

We had experienced an octopus explosion. As they grew hotter the tentacles had stiffened and forced their way out of the pot in one gigantic reflex action.

Exploding octopus: with salt, pepper & onions.

We discovered this day that, although difficult, you can get diluted octopus ink off walls, floors and ceilings with washing-up liquid and lots of effort. The worst stains were the ones on the ceiling, but they eventually came off also. Noel threw the curtains into the washing machine and that seemed to do the trick, but we heard later that his mother asked why the curtains looked 'different' because his ironing skills were not the best. Long after, he told them what had happened. His mother was a bit upset but Noel said his father had a good laugh over it. Mr Berry was actually a really good guy, with a great sense of humour once you got to know him. He was a lot older than all our parents because they had Noel later in life, but I got along with both of them. I think they were both saints.

The crabs of Fairyland Creek had a feast of partially cooked octopus that day.
 I wondered if they appreciated the delicate seasoning of onion, salt and pepper.

As a footnote to this octopus story, many years later I was fortunate enough to work at the Bermuda Aquarium, Museum and Zoo as their Merchandising and Concessions Manager. This meant that I had the opportunity to meet and observe the dedicated staff at BAMZ. They were always ready to talk about the creatures in their care. One of the amazing things I heard about was the intelligence of the octopus. I went along one day to watch a demonstration that was being staged for a group of school children. Standing in front of the octopus tank we all watched as one of the aquarists placed a screw-top jar into the tank. Inside the jar was a smaller jar. Inside this jar was a juicy shrimp. Octopuses love shellfish. Both jars had their lids screwed on tight. As we all watched, the octopus approached the jars resting on the bottom of its tank. It seemed to glide along the sand and it reached out with a few tentacles and seemed to be assessing the situation. It slowly climbed all over the jars, delicately examining the outer jar. It then started to test the top of the outer jar. The aquarist had told us that he had put the top on very tightly. Slowly but surely the top started to turn and soon it was off. The rest was over in seconds.

Now that the octopus knew the solution, it was a simple matter for it to reach into the large jar and pull the smaller one out, unscrew the top and pull out the shrimp and eat it.

After the demonstration, I asked Patrick the aquarist if octopuses were really clever and he said that they were. Some time ago they had an octopus in one of the tanks and now and then a fish or crab would go missing from an adjoining tank, sometimes two or three tanks away. The octopus was climbing up the tank wall and into the one-inch water inlet pipe at night, crawling along and into the other tanks and exploring them, looking for a snack. It then had the intelligence to climb back into the pipe and find its way home. They solved the problem by installing plastic grass along the upper walls of the tank around the inlet pipe and it didn't like the feel of it and stopped climbing out of its tank. As long as they can fit their beak through an opening, they can get the rest of their body through. They are truly amazing creatures.

MOPED 3 - Zundapp

In 1958 the bike of choice was a Zundapp. If it sounds German, that's because it was German. They were nice looking bikes, a metallic blue-green and although still only allowed a 49cc engine, they were quite nippy. Unlike my previous mopeds, they even looked a bit more like a motorcycle with the larger tank than the Mobylette, and they went faster as well!

The "dapp"

THE PARTY

It was 1958 and Noel, back on the island for his Easter break, had decided to work at the Bermudiana Hotel as a pool attendant. I think his parents always felt disappointed that they were spending all sorts of money on his education only for him to get jobs that were not suitable for the executive son that they longed for. As far as Noel, and the rest of us, was concerned, working as a pool attendant at the hotel was a stroke of genius. In those days Bermuda was a major destination for college students in the northeast USA during their 'Spring Break'. A lot of them stayed at the Bermudiana Hotel in Hamilton. Noel spent all day offering towels and cold drinks to hundreds of nubile American college girls. Naturally quite a few of them received his attention in the evenings as well. He was quite a good-looking fellow with curly brown hair, twinkling blue eyes and a way with words. His chosen subjects in college of English Language and Literature were not completely wasted. I am sure his parents would have been grateful.

Later that year Noel came home for another Summer break and his mother mentioned to me that she wanted to organise a nice party for Noel for his eighteenth birthday. It would be on Thursday, September 4th. It was ideal because Thursday was a half-day in Bermuda for most businesses and she wanted to start the party early so it didn't last until late at night. Noel was still working at the Bermudiana, much to his father's disgust, and I had made some fake plans to do something with Noel so he would be surprised by the party. There would be a dozen or so people coming, guys and gals. His mother had left me in charge of the invitations.

We had made plans that I would meet him at the hotel and we went to his house where some people were already showing up and he was pleasantly surprised. His mother had gone all out as she considered it an important birthday. So did Noel because it meant that he could drive his father's car!

Noel acted surprised when we got home, although I think he had suspected something was up, and as people started to drift in, we had the record player on and listening to the latest sounds of the day. Looking back on it now, in those days, we had some great music with some good songs with great beats and words you could actually understand, unlike the songs I hear on the radio today. I believe that's what they call the generation gap. The top sounds then were 'Bird Dog' by the Everly Brothers, 'Yakety Yak' by The Coasters,

'Great Balls of Fire' by the wild man Jerry Lee Lewis and 'Hard Headed Woman' by the big name of the year, Elvis Presley. We were sitting around, some of us having a cold beer that Mr Berry had provided. Everything was looking like there was going to be a good party and Mrs Berry was delighted.

At about 6:00 pm One-Shot arrived at the party, all excited, and said that he had just heard that the Bermudiana Hotel was on fire! Nobody believed him, especially Noel and I who had just left the hotel not so long ago. We turned on the radio and then we heard the bulletins for ourselves.

The excited announcer informed us that the fire had started in one of the top floors and was slowly getting bigger. The hotel guests had been asked to evacuate immediately. The new Bermuda television company ZBM-TV had just started up in January and they had offices right across the road from the hotel. We turned the television on and watched the black-and-white pictures. Any idea of a birthday party vanished.

"We should go down there and see it," said Noel.

The look on Mrs Berry's face said it all. She had worked hard on the sandwiches and cake.

"You can see it all on the television."

"It's not the same, mum. We won't stay long, I promise."

Famous last words.

It was too much to expect us to stay. This was the biggest thing to happen in Bermuda for years. A hotel was on fire just down the road from where we were standing. Before long we had all hopped on mopeds and tore off down the road, leaving Noel's mother standing at the door and his father glowering at us through the kitchen window. We all raced down to the crossroads of Fairyland Road and Pitts Bay Road and turned right, heading into Hamilton and the fire. We didn't have any real plan because we didn't know what to expect. At the time I think we thought we could just arrive at the scene and find a vantage point and watch all the excitement.

When we arrived at the junction of Woodbourne Avenue which bordered the hotel and led up to the entrance, we were confronted by a very stern-looking policeman.

"Where do you boys think you're going? Don't you know there's a fire?"

We retreated and huddled to discuss Plan B...which didn't exist, but we soon had one. As usual, it was Noel who came up with a plan. After all, he worked at the hotel.

"We'll go back along Pitts Bay and up Rosemont then along to Richmond and come in on Woodbourne from the back. We can get into the back of the hotel grounds by the cycle shop from Gorham Road."

Off we went as fast as our 49cc mopeds could take us and finally we squeezed through the oleander hedge along Gorham Road at the back of the cycle rental shed behind the hotel. There were some discarded boxes and shipping pallets behind the shed and before you could blink an eye there were six of us lying along the top of the flat roof of the shed with a bird's eye view of the hotel.

By now it was almost 7:00 pm and the fire had grown from a small trickle of smoke to a blaze on the upper floor. Between us and the hotel was the swimming pool, and amazingly somebody was swimming in it.

"I don't believe it! There's somebody swimming in the pool."

"I guess they haven't heard all the fuss."

"They must be deaf and blind."

The Bermudiana was a large hotel by Bermuda standards and although quite nice, it was over 30 years old and needed some renovation work. It looked like it was going to need a lot more after tonight.

Although we could not see all that was happening from where we were, we heard later that when the firemen arrive on the scene the first thing they found was that there were no fire hydrants. In fact, the Hamilton Fire Brigade did not have adequate equipment to fight a fire in a building of this size.

The first attempt to spray water onto the fire resulted in a weak trickle of water that fell onto the lawn and flowers of the hotel. Later on, when it was too late, they managed to get the water onto the fire but it was hopeless.

From our vantage point, we could see things being thrown out of the hotel windows. Some of the guests took the time to pack their belongings into a suitcase and now and then a suitcase would be pushed out of a window to fall onto the ground below. Other guests were simply throwing their clothes onto the bedsheet, tying it all up into a bundle and heaving it out of the window. We saw several of these come flying down to the ground. There was a dentist's office in the hotel and some equipment, including a dentist chair, came out of that window!

As it grew dark the fire glowed a bright orange-red against the darkening sky. We could feel the heat even though we were quite a distance away from the building. As it became clear that there was nothing they could do to save it, we climbed down from the top of the shed and got on our bikes to head back to Noel's house. What had started as an adventure had turned into a sad spectacle of the destruction of a Bermuda landmark. We got back around 10:00 pm and we told Mrs Berry we would come back and enjoy her sandwiches and cake on Friday night.

The next evening we all gathered at Noel's house, and it was a very good evening because we had so much more to talk about!

MOPED 4 – Motom

The Zundapp didn't last long because a new bike was top of the list of most wanted bikes.

If it sounds like we changed mopeds almost as often as I changed my socks, it's true. Peer pressure is a powerful force, and as soon as one member of our group changed to a bike, it was almost certain that we would all follow, and as soon as the Motom arrived on the island, they were a big hit.

There was so much to like about the Motom. First of all, they LOOKED like a motorbike, with the big tank just in front of the seat. But the best thing was the sound. They had 4-stroke engines which made them sound completely different from the old 2-stroke mopeds we were used to riding. If you fitted them with the rear section from a Triumph exhaust, they really sounded great! The other attraction was that they could do forty miles an hour! This might not sound like much, but on an island where the speed limit was 20 mph, it was wonderful. The police weren't too happy about it, because they now had a plague of speeding teenagers to contend with on their 125cc police bikes.

We grew to know some of the police officers who weren't bad fellows, just doing their jobs. The ones that were quite friendly would stop and have chats with us. We would tease them and ask them how many green tickets they had handed out that day. Green Tickets were little printed warnings on green paper that were handed out to tourists if the police officer thought their attire was too revealing. The ticket could be given to men or ladies, but the police gave them mostly to ladies who wore swimsuits in the town. Bermuda was quite puritan in those days.

The exact wording on the Green Ticket was :

"May we respectfully suggest that your attire may prove to be embarrassing as there are certain regulations pertaining to propriety of dress that are being enforced in order to maintain Bermuda's position as a most attractive and pleasant holiday resort."

The general opinion among the Bermuda populace was that a lot of the policemen, many of them young men from Great Britain, found that the Green Ticket was a very useful way to meet attractive young lady tourists!

The only problem with the Motom was that after riding it flat out at top speed for any length of time, or when you raced somebody on one of the few stretches of roads suitable for racing in Bermuda, the tappets needed to be reset. Needless to say, we became quite expert at adjusting those tappets.

To enable me to purchase the 'must-have' Motom, I had put out the word that my Zundapp was for sale to the first reasonable bidder to contact me. I was soon contacted by a character called Rocky and it was all very clandestine. We agreed to meet at the parking lot of the Paget Pharmacy and I went there with Noel so he could give me a ride back home. Rocky showed up with one of his buddies and they were very tough-looking guys. He had the required cash in his hand and I had made up a very basic Bill of Sale and once the transaction was completed we parted ways. As they rode away, Noel and I looked at each other and grinned.

"Man, those were some hard-looking guys."

"Yeah, I wouldn't want to meet them in a dark alley some night."

"You're right. No way!"

A week later I was riding my new Motom past the same pharmacy and going up Strawberry Hill that curved up and around a rather sharp corner before descending down and into Hamilton.

Above the noise of my new Motom I could hear a dreadful clatter coming from around the corner, up ahead of me. The next minute, coming toward me, was my old Zundapp with a wild-looking Rocky hanging on as if his life depended on it. There was no tyre on the front wheel of the bike and the steel rim was making a terrible din as the bike raced along and with no rubber tyre on it, the bare wheel was skidding sideways on the hard tarmac. I slowed down as he approached and watched in awe as he flashed past me, the bike actually skidding, rattling and bouncing around the corner, with Rocky, his eyes wild with excitement, grimly hanging onto the handlebars. A bike with two fellows on it was following him and they were laughing their heads off. I continued on my way and never saw him again, although I heard that he had survived his crazy ride. It seems that one of his equally crazy friends had bet him that he couldn't ride the bike with no tyre on it, so he had taken the front tyre off and driven down the hill. I assume he won his bet.

The Motom

THE WEDDING

It was One-Shot who told us as we all gathered at the dock house, "Hey you b'ys. My cousin Eddie told me his cousin Tony is getting married to his girl Maria over at St.Anthony's tonight."

In 1959 the population of Bermuda was only approximately 40,000. If you put all our little gang of guys together, it was quite likely that we knew, or knew somebody who knew almost half the population. It was a fair bet, therefore, that if somebody was getting married, we would know somebody in the wedding party.

"Is that Joe's sister, Maria?"

"Yeah, that's the one, their father has the vegetable market on East Reid."

" Oh, right, her father is my auntie's cousin."

The conversation went like this for a few minutes and then Noel said,
 "We should go to it."

Ronnie spoke up, "Yeah, we could go, I think that's where my parents are going tonight."

"I'll see if I can get my dad's car tonight," said Noel

"Could we all get in your car?" asked One-Shot.

"Well, Florence was coming over, so I guess I should take her with us," said Noel.
Florence was Noel's girlfriend.

"Some of us can go on bikes, I guess," said Ronnie. Florence would get the front seat. We all knew that was the rule.

Later that evening a group of us converged on St.Anthony's church in Warwick, Bermuda. We had all put on our best dress for the occasion. None of us actually owned a suit. Our formal wear consisted of a pair of slacks, hurriedly pressed, some sort of jacket, a white shirt and a tie that was probably the only one we owned. We even had put on socks which were a huge decision in our little gang. We hardly ever wore socks.

St.Anthony's is a nice little church set back slightly from the Middle Road, built on a hillside that sloped down to a large level playing field next to White's Market. The church had a community hall beneath it and the reception line was set up there, along with the buffet and bar. We parked on the field along with dozens of other small cars. Bermuda regulations strictly controlled the size of cars that could be sold. Mr Berry had a Morris Shooting Brake estate car. It had two doors and varnished strips of wood on the sides. It was considered very smart although not much bigger than a Mini-countryman of today. The main advantage of the car was that it was quite roomy for its size and had two half doors at the rear that gave easy access to the area behind the rear seats.

There was a small grassy terrace immediately outside the community hall where the bar was and then steps down to the field where a large number of tables and chairs had been set up. Cars were parked outside this area, near the access to Middle Road. When we arrived the wedding ceremony was finished and everybody was slowly shuffling along in the reception line wishing the new couple a long and happy life together. It was a Portuguese wedding so there were lots of people there, and I got in line with One-Shot who knew the groom's aunty, or so he said. I never knew with One-Shot if he was telling little white lies or not. Not that he was dishonest, but he could stretch the truth to suit the occasion.

As we stood in line there was lots of jostling and poking in the ribs with elbows. We all thought it was a big joke. We got to the bride and groom and the groom looked at One-Shot, then at me with a funny little smile, as if to say, "Who are these guys, I don't remember inviting them."

With great sincerity, One-Shot said "Hey Tony, remember me, Roddy DeSilva? I want to wish you all the best, mate. This here is my third cousin, visiting from the Azores," as he pointed back at me with his thumb and then shook the guy's hand. "Maria, you look beautiful, girl!" as he took the bride's hand and gave her a kiss on the cheek. One-Shot was never short for words, and I mumbled a few pleasantries, trying not to sound too tongue-tied and rushed after One-Shot who was headed toward the bar. Behind us, the same scene was being played out by the rest of our group, with the ones who vaguely knew the happy couple vouching for the others.

One-Shot's gambit of introducing me as a visiting cousin from the Azores was not as outrageous as it might sound. The large Portuguese population of Bermuda had mostly originated from the Azores so it would not have seemed strange to Tony, except for the fact that I probably looked very lily-white compared to most Portuguese.

We went to a table down on the field and then as the rest of our gate-crashers joined us we got more and more chairs around it until there were eight of us sitting in a happy group. There was a small group playing dance music and there were lots of young ladies at the wedding and as you can imagine, with a bunch of handsome debonair lads like us, we soon had our own little party going.

Quite soon Noel got up and said, "Time for a refill," and walked up the steps to the bar. He soon came back with two cold Heineken beer bottles in each hand. He casually walked past our table, placed one bottle on it and kept going straight to the car where he quickly opened the rear doors and slipped his hand inside and then shut the door. When I had joined Noel at his house to come to the party in the car, he had disappeared into the cellar for a minute and returned carrying two cardboard boxes and thrown them into the back. Now I understood.

After a few minutes, first Ronnie, then One-Shot and all the others in our group did the same thing. "Time for a refill," and they would disappear toward the bar and return with four beers, place one on the table and head for the car. It was my turn so I walked up to the bar, looked the bartender in the eye and said, "I need four Heinekens for our table, please" The bartender never even blinked. He just popped off the tops and placed them in front of me and turned to the next customer. I walked to the table, put one bottle down and went to the car. I opened the door and inside were two cardboard Heineken boxes with the little partitions inside them to accommodate a dozen beer bottles each. When I placed my bottles inside the second box, it was full. Two dozen cold beers. I walked back to the table and sat down. "Must be full by now, right?" asked Noel. The others looked at me with expectant grins. "Yep, all full," I said and grinned.

"Well, I think this party is coming to an end here," said Noel. "Time to head for the dock shed."

With that, we all got up and headed for our transport and went to Noel's house where we had a party, courtesy of the happy couple Tony and Maria.

I hope they are still happily married.

JAWS

"I must have lost a dozen brand new balls in there," said Mr Berry.

We were sitting in the Berrys' living room. Me, Noel and Mr Berry, on one of the rare occasions when Noel sat down with his father and had a real conversation with him. Mrs Berry was visiting a friend and we had thought he looked a bit lonely. Normally Noel would have headed for his room or the dock house, but now we sat and chatted with his dad. Noel had casually asked his father how his golf game was going. Neither of us was interested in golf in the slightest, but it seemed a safe topic of conversation. Noel seemed to perk up at this piece of information.

"What hole is that, Dad?" His father was a member of Riddells Bay Golf Club.

"That bloody Hole 9 across the water," said his father.

"Why do you lose balls there?"

"I know I shouldn't, but I think I am intimidated by the water and I tighten up and try too hard and end up topping it and it just gets half-way across and splashes into the little bay. On a calm day you can look down and see dozens of balls on the bottom."

"Oh really?" said Noel, glancing at me. I was trying to follow the direction the conversation was going. Noel was not interested in golf, so why was he showing so much interest in his father's golf problems?

"Where is this hole, dad?"

"It's right at the western tip of the golf course, the place called Burgess Point."

"Oh yeah, I know where you mean. Well, dad, I guess you need to simply relax and hit it as if the water isn't there. Maybe that will help."

"I hope so, those new balls aren't cheap. There must be a hundred of them in that little bay," his father moaned.

A little light bulb went off in my head, a hundred expensive almost-new golf balls lying in a little bay? Was Noel already thinking what I was now thinking?

Ker-ching!

The next day Noel was still working at the boat rental place, and business was a bit slow because it was late summer and the tourist season was slowing down. It was a Saturday so a few of us were off work that day, and four of us jumped into one of the rental boats and headed for Burgess Point.

There was me, Noel, Ronnie and One-Shot. We were geared up as if going on a dive on the Titanic. In the bow of the boat there were dive masks, snorkels, a couple of pairs of flippers, some rope and an anchor and some little net bags that Noel had 'borrowed' from his mother's laundry basket in the cellar.

"What does she use those for?" I asked.

"I have no idea," he replied. "But they should come in handy today."

I wondered what his mother would think if she knew her laundry accessories were going on a diving expedition.

As we glided out of Mills Creek and headed out through the islands at its entrance and then past Two-Rock Passage where the ocean liners threaded their way into Hamilton Harbour, Noel outlined his plan.

"My dad says that there are hundreds of almost new golf balls in this bay at Riddells Bay golf course. It should be easy for us to go in there and pick them up, and then we can set up outside the entrance to the club and sell the balls back to the members. "

"Cool!" said One-Shot. "Maybe we'll be selling them back to the golfers that lost them."

This raised a laugh around the boat. The thought of selling them their own balls was hilarious.

"I wonder if they could recognise them. They're numbered aren't they?"

"Yeah," said Noel. "My dad says that some of the members actually get their balls personalised with their initials."

"No shit?" said Ronnie. "Man, that is really showing off."

By now we were rounding Hawkins Island and heading past Grace Island on our right, with Burgess Point straight ahead. It was a lovely autumn day in Bermuda, not too hot, calm with a slight breeze from the southwest, a perfect day for snorkelling in a shallow bay.

We rounded Burgess Point and slowly drifted past the larger sandy bay below it, and then reached our destination.

The bay was about 50 yards across and about 75 yards deep and we could see a few golfers in the area, but none were actually playing across the bay. As we idled around the little rocky island at the entrance to the bay, we could already see little white balls resting on the bottom amongst the seagrass that covered the bottom of the bay.

"We better anchor here at the entrance, it looks pretty shallow. I think we could almost stand on the bottom a bit further in," announced our captain.

Ronnie threw the anchor over and tugged at it to make sure it was holding. There was hardly any wind so there was no chance of the boat drifting away from the bay, sheltered as it was by the little island.

"Right, let's go make some easy money," I said. Everybody stripped off their shirts and put on dive masks and snorkels. Nobody bothered with flippers. The bay was far too shallow for them to be useful. All of us holding one of Mrs Berry's net bags, we jumped in.

Each of us went in a different direction. I went south and the others went in closer to the shore. It was brilliant. The sun filtered through the water to shed light spots across the bottom of the bay among the seagrass, which was about a foot high. The water was warm in the shallow bay where it had been heated by the sun for hours. I glided over the surface of the bay looking down and it wasn't long before I spotted golf balls. Mr Berry was quite right; there must have been hundreds of them. I did a few shallow dives and quickly recovered five or six balls in quick succession. I popped my head up to see legs sticking up out of the water as the others reached down to pick up balls.

It was so shallow that it wasn't right to call it diving, I was simply bending over and picking them up. The water was crystal clear and the whole exercise became very relaxing as I drifted along scooping up all the lovely balls that were going to make us rich.

I was still on the western edge between the boat and the southern shore of the bay. I could see the others closer to the shallow part of the bay, scattered around so as not to interfere with each other. I put my head underwater to look for more balls.

Something caught my eye, something that I didn't expect...something that shouldn't be there. I had just plucked a nice clean golf ball from the seagrass and placed it into my net bag, pulled the drawstring tight at the top and floated stationary in the water, slowly looking around.

There it was. We were recovering golf balls from a shallow grassy bay. The water was nice and warm. Barracuda love warm shallow grassy bays. There was one immediately in front of me. He must have been four feet long. He was sideways to me and looked at me out of one eye. I looked back at him. He didn't move, didn't swim away, he just looked at me, his mouth slightly open and showing an alarming number of very sharp teeth.. He seemed to be saying, "This is my territory, buster, and if anybody is going to move, it's you!"

He was only about five feet away. I suddenly remembered being on camping trips out at Daniel's Head with the scouts and seeing schools of barracuda off the rocks there. We used to throw the leftovers from our suppers to them just to see them feed. Now I remembered... they travel in schools.

Very slowly and trying to keep calm and not getting too alarmed, I turned my head to the right. There were two more of them, the same size. I turned my head to the left, and 'Oh shit!' there were three more.

The scariest thing about barracuda when you're in the water with them is that they don't do very much except watch you. If you are diving, they just follow you, up and down or back and forth. Seasoned divers say that they are just watching you to see if you stir up anything for them to feed on, but some also say that if they see something shiny like a ring, bracelet or watch that catches the sun, they will think it is a small fish and come after it. The fact that it is attached to you doesn't seem to bother them.

I didn't own a water-proof diving watch and I didn't wear rings or bracelets, but that fact didn't stop me from quickly checking my arms and hands to make sure I hadn't absent-mindedly put something shiny on without realising it that morning.

"OK, nothing shiny here, Mr Barracuda...no sir, nothing to interest you!" I murmured to myself, and also to the big fish nearest me, just in case he was listening.

They all just looked at me. I was getting decidedly uncomfortable. I slowly lifted my head to see where everybody was. Then I had another thought. Were there any more barracuda? Very slowly I looked behind. Thank goodness, they were all in front of me. I looked at the rocky shore, not too far away.

I trod water, took my snorkel out of my mouth and yelled at the top of my voice, "BARRACUDAAAAAAAAAAA!" and in the same instant, I did a pretty damn good imitation of an Olympic swimmer going for the gold in the 50-yard event.

It couldn't have taken me more than five seconds before I was scrambling up on the rocks, but each second I swam I was expecting to feel a bite on my feet. I stood on the rocks and looked around. Three bodies popped out of the water and onto the rocks around the small bay.

"Barracuda?"

"Where?"

"How many?"

"How big?"

"Six big bastards!" I yelled across the water.

"Oh, man!"

I pointed at the big fish but they couldn't see the fish because of the sun's reflection on the water, but I could see them quite clearly.

Noel spoke, and my heart sank.

"You're the closest to the boat. You have to get to it and bring it in closer to us."

"Who, me?"

"Yes, you." He was grinning, the bastard.

"They won't hurt you, man," this from Ronnie, who had done quite a bit of diving. The trouble was, he was grinning too.

"Well, if it's so safe, why don't you all just swim out to the boat as well?"

"Point to where they are."

I pointed to the big fish, and it was clear they were between the boat and the rest of the guys, but not between me and the boat.

"Go for it, man," said Noel.

Easy for him to say, he wasn't going to swim past a bunch of ravenous man-eating giant barracuda, at least that how it seemed from my point of view.

Tucking my bag of golf balls into my trunks, I took a leap of faith and did another Olympic swim to the boat, breaking the world record and hauling myself on board as if a school of barracuda was nipping at my heels. A far as I knew, they probably were! I pulled the anchor off the bottom and, leaving the outboard engine in the raised position, I used one of the oars to paddle in close to the shore. My brave friends quickly waded to the boat and jumped on board. As I paddled out to deeper water, we all looked down.

"Damn, you weren't joking, were you?"

"They are pretty big!"

"What happened, you just sort of swam into them?"

"Yup, that's about the size of it. I looked up and they were all around me!"

"I bet you almost pooped your pants. Haha!"

With friends like this, who needs enemies?

We cruised back to Noel's dock, counting our golf balls. We had about fifty between us. We could have doubled that if it hadn't been for our fishy friends. The next weekend, when it was a busy time at the golf course, we set up a little table at the entrance and made a tidy profit. Naturally, it was quickly spent on beer.

++++++

The Great Barracuda

"A *barracuda* is a large, predatory ray-finned fish known for its fearsome appearance and ferocious behaviour." *Wikipedia*

"Swimmers have reported being bitten by barracuda, but such incidents are rare and possibly caused by poor visibility. Barracudas may mistake objects that glint and shine for prey. Barracuda attacks on humans are rare; bites can result in lacerations and the loss of some tissue." *Fish facts*

Not very reassuring!

MOPED 5 – CYRUS

The next, and last moped, was the Dutch Cyrus. Although the Motom was a cool bike, it was a pain in the neck because you always needed to reset the valves. The Cyrus arrived around 1959 and was immediately popular because it was fairly fast and very low maintenance. They became the mainstay of the tourist moped rental outlets. Everywhere you looked there was a Cyrus.

There was another reason that the Cyrus moped was an instant hit with the locals in Bermuda. It was manufactured as a two –gear bike, but Bermuda regulations did not allow mopeds to have gears. They had to be 50cc or less and be a single gear bike. The bike importers of Bermuda got around this technicality by having the lower gear, or first gear blocked off at the factory before being shipped to Bermuda. With a large rear sprocket on the back wheel, the bike still had plenty of power for riding around Bermuda's roads at 20 mph.

It did not take long for word to get around among the local lads that there was another gear inside the Cyrus engine and soon everybody knew how to modify a bike to make it into a two-gear machine. All you had to do was drill out the little pin that had been inserted through the top of the engine case to prevent the gear arm from engaging the first gear.

There was only one problem...it was illegal.

The solution was simple. We loosened the pin until it came out, then we drilled a tiny hole in the pin so that we could insert a piece of nylon fishing line or catgut through it to stop it from dropping down and blocking the first gear. The nylon line ran up the frame from the pin to the handlebars and was tied there. If a policeman stopped you to check if your bike was legal, you pulled the nylon line and the pin dropped into the engine and suddenly you had a legal one gear bike. What master criminals we were!

The Dutch CYRUS – great bike!

There was only one problem with modifying your Cyrus to make it a two-gear bike, and that was that it had lots of power for impressive starts but no real speed, but more about that later.

The police Bike Squad soon learned about our diabolically clever trick of pulling the nylon fishing line out of the pin to make the bike legal, and you had to be extremely quick at pulling it out and getting rid of the incriminating evidence before one of the policemen pulled you over. Soon there were lots of small lengths of discarded nylon fishing line scattered along Bermuda's roadsides that were testimony to the vigilance of the Bike Squad and our attempts to avoid being caught.

WHITE'S ISLAND

White's Island is a small island of situated right in the middle of Hamilton Harbour. As far as I can recall it has always had a house and large dock with changing rooms and a big hall built above the dock with windows on three sides with storage beneath. In its early years, the main building had been a large sail loft, then a part of the United States Navy and later it had been a casino. In the years that I went there to be looked after it had been renovated by the Bermuda Sailing Club.

Soon after I arrived in Bermuda at the age of nine my mother made arrangements for me to go to White's Island every Saturday so that the family living on the island could keep an eye on me while she went to work.
 The Lawrence family were great. They had several children and Mr Lawrence was a diver. He had a great big old brass diver's helmet on display and it was indeed a thing of great fascination for me. As far as I can remember there were two boys my age or a little older and a daughter called Cathy. I thought Cathy was beautiful, and she was the first person I ever had a crush on. It almost got me killed.

Every Saturday morning my mother would drop me off at the little dock near the flagpole on the Harbour and she would see me safely onto the old rowing boat that was manned by a lovely old, weather-beaten black gentleman who looked like he had been providing a ferry service across the harbour for centuries. The fee for being rowed to White's Island was threepence, and I can remember proudly handing my silver little 'thruppence' to the old gentleman who would treat the transaction with great seriousness as he carefully pocketed my treasure in his trouser pocket and pushed away from the dock. He would carefully hold the rowboat steady for me as I climbed onto the steps of the dock on White's Island and politely wish me a 'Good Day, young sir' after confirming that he would return for me at precisely five o'clock so that my mother could pick me up on her way home from work.

Once on the island, I would join the Lawrence children in games and eating lunch and sometimes helping Mr or Mrs Lawrence with light tasks that were allotted to us to keep us out of mischief, but most of the time was spent running around and playing. The dock on White's Island was quite large, at least to a small boy just arrived from England it seemed large. We played on it all the time and the Lawrence children used to run and dive off the dock and have great fun swimming and playing in the clear, warm water. I would stand back and look on enviously and with some feeling of self-consciousness because I could not swim. As I mentioned earlier, I thought that Cathy was lovely. She was always nice to me and had long dark hair and I would watch her as she ran, dived and swam with her brothers. I was not long before the devilish brothers realised that I was spending a lot of time watching their sister, and started teasing me. One day the boys were jumping off the dock and I was watching them when I looked up and saw Cathy coming along the dock to join us. The two boys looked toward their sister, looked at me and started grinning and saying, "Jump in John and show Cathy how you can swim."

To this day I have no idea what made me do it, but I jumped into the water, right off the dock. It was about four or five feet to the water and after I jumped I remember wondering what on earth I had done. All I knew was that Cathy was watching and I had to do it. You have to remember that I was newly arrived in Bermuda and had no swimming lessons or knowledge of what was required once you hit the water. For instance, nobody had told me about taking a breath and closing your mouth. I did neither. As I lay on the bottom looking up through the water, I remember that my mouth was open and I could feel the stones on the bottom rubbing on my back. Looking up, I could see three young faces looking down at me. They must have started yelling because the next thing I knew a very strong arm had grabbed me and pulled me up out of the water and onto the dock where I coughed up rather large amounts of seawater, and Mr Lawrence was saying "What the hell were you thinking, boy?"

I was thinking, "I bet that got Cathy's attention."

Next week my swimming lessons started and I was soon swimming along with the others.

THE RACE

Many years after my near-drowning, in 1959, the island hosted dances in the summer months for teenagers, arranged by the sailing club. We all attended these dances, not because we liked to dance, but because it meant there were lots of young ladies gathered together in one spot every Friday night. The dances were chaperoned by adults who kept an eye on things and there was no booze allowed. Everybody was expected to dress respectably and behave themselves. There was a special ferry service to get us all to and from the island at scheduled times. It was great.

One night we all attended and there was the usual bunch of guys including me and Noel along with One-Shot, Gordy and the rest of the gang. We had chatted to some girls and been talked into dances by some of them. I had two left feet but could manage a respectable Jitterbug and Twist without making a complete fool of myself or destroying my partner's feet. The Twist was the hot new dance from America and everybody was trying it, resulting in some interesting variations and lots of hilarity.

When it was time to go we all got onto the ferry and took the short trip to the Hamilton Ferry terminal where we gathered on the dock, talked and laughed, joked with the

girls, made dates and then got on our bikes and started them up.
It was all a very normal Friday night. As we sat on our bikes revving them up a bit and thinking we were pretty cool, another bike drew up alongside us on the road.

I remember thinking that the rider looked a bit like one of the legendary American Indians Geronimo or Cochise, with a classic face out of the movies with an intense look in his eyes and a fierce expression on his face. He sat and stared at us, revving his bike. It had a really cool, very deep sound. I realised he had modified the muffler.

Glaring at us, he threw down a challenge. "You guys think you got fast bikes, huh?"

We could see where this conversation was going. We all exchanged looks. One-Shot broke the silence.

"Hey, Winston. What's happenin' man?"

The stranger looked at One-Shot, gave a small nod and just said "Shot."

So, his name was Winston.

He spoke again. "So? You guys think you're fast?"

Noel spoke first. "Yeah, we think we're pretty fast."

Winston threw down the challenge.

"Prove it, first one to reach Spanish Point Grocers."

"OK. Let's go!"

With that, the challenge was thrown down and picked up. Six of us lined up outside the ferry terminal and revved our bikes, somebody yelled "Go!" and away we went.
The road to Spanish Point went out of town past the Bermudiana hotel site, and then along the quiet residential area of Pitts Bay Road, climbing a slight hill past Pitt's Bay before swooping down and past the turn to Fairyland Road and Noel's house, then up around a sharp left-hand bend and hill to the grocery store. It would be about two miles in length.
As we got going we all crouched over the handlebars and cranked up to our top speed of about 40 mph. It was a very even race and I was up at the front next to the challenger that we now knew was named Winston. Noel was right with me. He was a fearless rider, sometimes quite foolhardy and I thought he would catch Winston and might beat him, but as I looked over at Winston I saw him lift his right leg and kick down. Suddenly the night was filled with a deep roar and I realised he had knocked his muffler back from the down-pipe somehow. His bike surged ahead by a couple of bike lengths and no matter how hard we tried, Noel and I could not catch him. The other guys were right behind us.

As we neared the sharp left bend that went up the hill to Spanish Point I eased back but could not believe my eyes as Winston lay his bike over and kept going full speed around the corner and pulled away from us even further.

"That guy's crazy," I thought."Even crazier than Noel!"

We all pulled up outside the grocery store that was closed for the night.

You guys ain't fast," Winston taunted us.

"Not when you knock off your muffler," I said. He just looked at me.

"I can beat you without knocking back my muffler," he laughed.

"OK, back down to St.John's Church this time."

Off we went again, mostly downhill, along the road that passed One-Shot's house and then levelled out before it got to the church. He beat us.

"One more time," said Noel. "Back to the Bermudiana."

We all knew that, by now, we were risking getting caught by the police for speeding. Residents would have been on their phones complaining about the idiots who were racing up and down the roads disturbing their night.

"Let's go!" said Winston and off we went again.

This time the route went back toward Spanish Point but then the road took a very sharp left-hand turn toward Hamilton.

We tore along. On a small bike with no helmets, 40 mph seemed very fast to us, especially when we usually had to stick to 20mph. As we approached the sharp left-hand bend in the road, I remembered seeing Winston refusing to ease off when we had raced to the grocery store.

I was right behind him. He was brightly lit by my headlight. I could see the sharp turn coming up ahead. Across the road, there was a steep drop of about six feet into a banana patch.

"He's not going to back off!" I thought as I started to ease my throttle back.

I watched in part admiration, part horror as he pulled away from me and leaned over to take the corner at an impossible speed.

He almost made it, but then he went across the road and hit the low curb, bounced high into the air, beautifully lit in my headlight and soared over the first couple of banana trees and then disappeared, bike with him, into the dense foliage.

The rest of us pulled up at the edge of the road, lights shining onto the banana treetops, looking for any sign of life or movement. There was nothing.

"Holy shit!"

"Did you see that crazy bastard?"

"Never seen anything like it!"

"I guess we better go find the body!"

Half of us went down the steep bank into the banana patch as the others stayed with their bike lights shining into the darkness in a feeble attempt to help us see.

Noel and I felt our way through the banana trees toward where we thought Winston might have landed. I was terrified we were going to find a mangled and dead body. Suddenly we heard a sound, half moan, half-laugh.

"Oh man, what a ride!"

We crept forward and there, looking as if he had decided to sit down for a rest, was Winston. He was sitting quite comfortably, leaning against a large banana tree. He looked up with a crazy grin.

"I won, right?"

"Yeah man, you won," said Noel.

"Are you OK?" I asked.

By this time we were joined by One-Shot.

"I found the bike," he said.

"Is my bike OK?" asked Winston.

"Yeah man, it looks OK, not bad at all, just the handlebars twisted a bit, should be easy to straighten them out."

"Good," he said and slowly got to his feet, slightly shaky. "I can ride it home."

He did ride it home. We helped him push the bike out of the banana patch and back onto the main road, straightened the handlebars, had a good laugh at his amazing survival, and off he rode.

Winston flying high

THE MOON

One funny moment happened one evening after we had all gathered in the shed on Noel's dock and there had been quite a few beers consumed. To tell the truth, there had been a hell of a lot of beer consumed, and some had consumed more than others.

We were all sitting around telling jokes and silly stories and things had begun to get quiet. In our group that evening was a guy called Bobby. He wasn't a regular member and only got together with us on the odd occasion. I don't think he drank a lot and on this night I think that he had drunk more than he was used to.

The shed was lit by two large lights with big green shades that hung from the ceiling beams on their electrical cords. While we sat around in the shed, Bobby had gradually become quieter and quieter and had finally lay back on the floor of the shed and gone to sleep. Maybe the term 'passed out' would be more accurate. After about fifteen minutes, we were amazed to see Bobby suddenly sit up straight, his eyes fixed on the bright white light above him and start yelling, "The moon! It's the moon!"

He then jumped up on to his feet, launched himself at the light as if he was trying to capture it, missed it completely, lost his balance and stumbled rapidly forward. The double doors out to the dock were wide open and Bobby kept going, arms waving around wildly, stumbling down the two steps onto the grass outside. As we watched, mouths wide open in shock, he kept going across the grass and then with a final yell "The moon!" he disappeared off the edge of the dock and with a mighty splash hit the water. For a minute we all sat there looking at each other in astonishment, then there was a mad scramble to the edge of the dock where several of us reached down and hauled a very bewildered Bobby up onto the dock. He was very indignant.

"Who threw me overboard?"

"Nobody. You jumped!"

"Like hell I did. Who did it?"

"Nobody, Bobby. You just started yelling about the moon and ran off the dock."

"I did?"

"Yes. Honest. It was really weird."

"I think I'll head home now."

"Good idea, Bobby. Take it easy, man."

I don't think he ever believed us and still thinks somebody threw him off the dock.

BIKE TESTING

After racing Winston and losing to him we became friends, frequently meeting at his house on a hill on the south side of Harrington Sound and overlooking the sound. It was an old Bermuda house and, being built on a hill, it had a huge cellar underneath it where Winston had set up a large workshop. This is where he worked on and modified his bike. Once we got to know him we realised that he was very clever and we started asking him to help us with modifying our bikes.

We hung out at Winston's quite a lot. I think one of the reasons was that he had a very large cellar under the house and we could all crowd into it and even bring our bikes inside to work on if the weather was bad. We spent many evenings just sitting around chatting and solving the world's problems.

Another reason we (or perhaps it was just me) liked to meet at Winston's house was that he had a large family, consisting of three younger brothers and more importantly, he had two sisters, both of them were extremely attractive. Lynn was older than us and therefore not interested in talking to such boring young men, but it did not stop me from admiring her whenever I saw her. She was (and still is) an absolute knockout, with long dark hair and quite beautiful. His other sister, Carmen, was also a very pretty girl, slightly younger than us, different from her older sister with lighter brown hair and I always tried to say 'Hello' to her whenever I was visiting Winston. His sisters certainly made a visit to his house much more pleasant, but not one of us ever tried to go out with either Lynn or Carmen, mainly because I think Winston would have killed us if we had!

Most of the time at the cellar was spent talking about our bikes and how we might be able to improve them, which actually meant coming up with ideas of how to make them go faster.

It didn't take long for somebody to come up with the bright idea that if we replaced the rear sprocket on the Cyrus with a smaller one we could go faster and, with the illegal first gear, still have the power to take off from a standing start as well as get over the steepest hills. With a little delicate drilling, the rear sprocket from a Motom could be modified and bolted to the rear hub of the Cyrus. After that, all that was required was the removal of a few chain links to get rid of any slack in the drive chain, and we were in business.

By pure chance, I happened to hear of somebody who had a Motom sprocket for sale and the next weekend we were at Winston's cellar working on modifying my Cyrus. I was quite excited to be the first in our group to have the new setup. Winston already had one on his bike and the speed he was getting was very impressive, hitting almost 50 mph, which was like breaking the sound barrier as far as we were concerned.

As we tightened the last nut on the Motom sprocket and adjusted the rear wheel to tighten the chain, I was all set to jump on my Cyrus and take it for a spin.

"How about letting me test it for you?" said Noel.

"What? Are you joking?" I replied.

He looked at me with a grin.

"Aw, come on, you will get to drive it all the time, and I don't know when I will get a Motom sprocket for my bike."

I hesitated, "Um, I don't know, I was looking forward to testing it."

"Come on, it will only be a short run down the road and back. Then she's all yours."

Reluctantly, I agreed that Noel could test my bike. We were good pals and I didn't mind. Well, not too much.

The three of us, me on Noel's bike, Winston on his, and Noel on mine, rode down the hill from Winston's house to Harrington Sound Road and waited until there was a break in the traffic. We didn't want any slowpokes holding up our speed test. After a few minutes, the road seemed clear and we headed east on Harrington Sound Road, Winston leading on his fast bike then Noel in front of me.

As we built up speed I could tell that my bike was now a lot faster than it had been. The road winds along the shore with a low sandstone wall between the road and the rocky shore a few yards below.

We came to a medium right-hand bend in the road where there was a break of about ten feet in the wall where it looked like somebody was making steps down to the shore below the road.

Noel was right in front of me and he bent low over the handlebars and opened up the throttle to take the corner at full speed. I think he might have been caught unawares by the difference in performance that the new sprocket had made, but he was not the sort of rider to back off anyhow.

Riding on the left, as was the law in Bermuda, Noel approached the opening in the wall at full speed. As I watched from behind he took the corner too wide and rode straight into the end of the low wall.

It all seemed to happen in slow motion. The bike went from about 40 mph to zero in an instant. Noel's body kept going. He sailed gracefully over the handlebars of MY bike, several feet in the air and landed in the middle of the road. My bike swung around into the road behind him and fell onto its side.

For a second I wasn't sure what I was most concerned about, my bike or Noel. Then common sense took over and I ran to where he lay in the road.

"Oh my god, mate! Are you OK?" I knelt beside him.

He looked up with blood coming from a scrape on his forehead and gave me a brave grin.

"Well, your bike definitely seems to be pretty fast, mate."

Then, as he started to move around and tried to get up, we realised that he was very shaken and had a broken wrist, which had happened when he put out his hand to try to soften his fall. Fortunately, his head had only scraped the road. He was extremely lucky.

Having made sure Noel was OK, I turned my attention to my beloved bike. It was a mess. The front forks had been pushed back into the engine and the wheel looked like a figure eight where it had slammed into the solid stone wall. The headlight was smashed and the handlebars were at the wrong angle.

It took a week to get my bike fixed, but Noel's wrist was in plaster for four weeks. To be fair, he paid to get my bike fixed up like new and I had lots of fun on it with my new sprocket, but I was stuck with ferrying him around while his wrist healed.

Never let your best friend test your bike.

WANT TO GO HITCH-HIKING?

"You want to go hitch-hiking around Europe?"

It was March 1960. We were sitting on the steps of the dock, each of us sipping a beer from the bottle. I turned to look at Noel as he asked me this strange question.

"You mean me and you?" I asked.

"Yeah, man. It would be cool. You and me, on the road together," said Noel.

"Are you serious?" I looked at him, not sure whether he was just day-dreaming or serious. We used to do this sometimes, imagine what it would be like if we could go riding Harleys around the USA, or go down to the West Indies and crew on a yacht.

"Yeah, man. I've been thinking about it. You know that yacht that's in Hamilton?"

I knew he had been getting to know the crew of the nice yacht that had been tied up in Hamilton for about a week.

"Yeah, what about it?"

"Well, they need a deckhand for the trip from here to Lisbon. I was talking to the captain and he seemed to think I could have the job if I want it."

"Man, that sounds cool! But what do you mean about hitch-hiking around Europe."

"Well, I was thinking if you came to Lisbon or Spain to meet me and we could hitch-hike all around Europe. It would be so cool. We could stay in youth hostels and it wouldn't cost very much."

I looked at him, making sure he was serious.

"You're serious, aren't you?"

He just grinned at me.

I thought about my job at American International. It was going well, I got on with everybody and I had been given some very nice pay rises over the last two years, but to be honest, I was getting restless going in and sitting at a desk, working nine to five every day.

"Around Europe, huh?"

"Yeah, man. It will be so cool." I think he knew he had me hooked.

"How soon?" I was thinking about how much notice I had to give to get any vacation pay coming to me. I liked my boss Mr Buswell and wanted to do the right thing and give him enough notice too.

"Well, they are talking about leaving in a couple of weeks and they are going via the Azores then to Lisbon so I guess we would get to Lisbon around the middle of April.

I thought about it. That would be another month. I could give plenty of notice to my boss, and there would be time to get organised.

"What about money?"

"We would hitch-hike everywhere. Some kids in Upper Canada college have done it and they told me it is really easy. As long as you are clean and shaved and don't look like a weirdo serial killer, lots of people will give you a ride. I know lots of guys and girls that did it."

"Where would we sleep?"

"They have these things called youth hostels all over Europe where you can stay for next to nothing. It's not The Princess Hotel, but it's clean and safe."

After that day, I started to think seriously about the idea, with Noel constantly pestering me to commit to it. There was so much to think about. How long would we be gone? He didn't know.

"It would be for months, maybe longer. We just see how it goes, man. Two free spirits, on the road together."

It sounded like The Great Adventure every young man dreams about. I floated the idea past my boss and he was great. He said he wouldn't stand in my way if it was really what I wanted. I would have quite a lot of money. With Christmas bonuses and savings I had put away a nice little sum of money in my bank account, and there was only one thing that made me hesitate.

One day while walking around Hamilton, I had walked past the car showroom of the local dealer for Austin cars and had suddenly stopped dead in my tracks. I looked through the large window into the showroom at the most beautiful car I had ever seen in Bermuda. It was a bright green Austin Healy Frog-eye Sprite. It was an utterly ridiculous sports car for driving around Bermuda with its narrow roads and 20 mph speed limit but I wanted it.

I had almost saved up enough for the down-payment and even some extra cash for a bit more besides. Now Noel had thrown a spanner in the works with his suggestion of a trip around Europe. I had to decide between the green Sprite or Europe.

In the end, the desire to travel around Europe with my best friend won. After much discussion, we decided on a plan... sort of...

THE PLAN

"It's simple," said Noel. "I'll sail over to Lisbon on the yacht and you can fly there and meet me. Then we will hitch-hike all over Europe. It'll be great!"

"When?"

"What?"

"When! When will you sail? When will you get there? When should I fly from Bermuda? "

"I don't know! I haven't worked all that out yet."

I looked at him. He obviously hadn't thought it through.

"What about money? How much will we need? Do we take cash?"

"My dad suggested we get traveller checks from the bank."

In those days every teenager didn't go around with their own personal credit card.

"Okay, but I need to give my boss a month's notice because that's how we're paid. It's almost the middle of March, so the soonest I can get away is mid-April," I said.

"Well, the captain said he wanted to set off by the 1st of April, so that should work out."

"OK, but what about the youth hostels, how would be able to stay in them. Don't you have to join up or something?"

"There's a book in the library about them. I'll get it and find out how we join."

"Yeah, it would be nice to have somewhere to sleep once we get there."

That's how it went for the next few days. It slowly came together. I should point out that, back in those days, we couldn't just go onto a computer and look up 'youth hostels' or 'hitch-hiking in Europe'. To plan things in the 1960s you had to do research using books and maps that you found in the library.

I approached my boss and he was sorry to hear that I was going. We agreed that April 15th was when I would work my last day at AICo, and the payroll department would pay me a few days in advance for any money due me, which would include the vacation pay that I had earned. This was all good news and cheered me up immensely. What didn't cheer me up was the fact that I still didn't know how I was going to get to Europe.

I had arrived in Bermuda early in 1948 on a BOAC flying boat out of Baltimore Harbour. All I could remember of that flight was that it was very long and very bumpy. (Many years later I discovered that I had been on one of the last commercial flying boats to operate into the island and it may very well have been the last one.) In 1948 there was no such thing as aircraft flying smoothly above the weather at incredibly fast speeds. The flying boats flew about 180 miles an hour at about 10,000 – 15,000 feet, compared to today's jet aircraft which fly at 600 mph at 40,000 feet. I can't remember the exact duration of my 1948 flight, but it would have taken about 5 hours to fly the 700 miles to Bermuda! A lot depended on the weather in those days. All you needed was some strong headwinds to add another hour on to the flight time.

So there I was in Bermuda, trying to plan a flight to Madrid and the last time I had flown or even travelled outside of Bermuda, for that matter, had been on an obsolete boat with wings stuck on it that barely got above the water and flew through wind, rain and thunderstorms to reach this gem of an island in its turquoise waters. Mark Twain, the famous author of Tom Sawyer, is quoted as saying that "Bermuda is a paradise but you have to go through hell to get there". Of course, he was talking about the many sea voyages he took to Bermuda, but I think the same could be said about those earlier flights as well.

My only outstanding memory of that last flight in 1948 was that it was long, bumpy and went on forever, and when we landed in Hamilton Harbour the passengers were taken into the Bermuda International Air Terminal on Darrell's Island. When we got inside and were going through the arrivals procedures, the lovely stewardess asked me if I wanted a cool drink. I immediately said 'Yes' and was a bit disappointed when she returned with a large glass of ice with milk added. The ice quickly melted and I was left with a glass of very watery ice cold milk. It's funny the things you remember.

I must admit that I was not very knowledgeable about travelling. I was a complete novice, and as I shopped around and asked anybody about travelling to Europe, they all agreed that I would have to fly to New York, then to London and then to Europe. When all these fares were added up, it became obvious that after paying for my flights, there wouldn't be very much left for the great hitch-hiking adventure.

Then one of the travel agents, a bit more expert than the others, said to me, "Why not fly on the Mexican airline, Guest?"

She went on to explain there was an airline called Guest Aerovias Mexico that flew through Bermuda about twice a week to Lisbon and Madrid. This was exactly what I was looking for. They flew from Mexico City to Miami, then up to Bermuda to drop off and pick up passengers, then on to Santa Maria in the Azores and Lisbon in Portugal and finally Madrid in Spain. Even now, writing this itinerary, my mouth waters at the thought of these destinations.

Realising I had the solution to how I was going to get to Europe, I sat down with Noel.

"When should I travel?" I asked.

He could have simply suggested a date, but instead, there was a new complication. "My father has suggested that we can stay in his friends' apartment in Milan, Italy if we want. They are going away for a month," He told me.

"Really? That sounds cool." (We actually talked like this in those days, come to think of it, I still do.)

"Yeah, he says it's a very nice apartment in the middle of Milan, which is a great place, according to his friends."

"So we just go to Milan and stay in the apartment? "

"Well, sort of."

"What do you mean, sort of?"

"They will be leaving on April 25th and would want us to get there a couple of days before that, to show us the apartment and how everything works, where to shop, that sort of thing."

"Oh, OK. So that means what?"

"We need to get there by April 23rd."

Now that we had a definite date to aim for we could begin making plans. "Is the yacht still leaving on April 1st?"

"Well, I checked with the captain yesterday, and we might leave a day or two earlier."

Now it was down to me, the guy that was supposed to be good with figures.

"How far is it to Lisbon and how fast does the boat go and when would you arrive in Lisbon?"

"I'll check with the captain."

This was going to be a long discussion, I could see that. The next day Noel had the information.

"It's about 2,300 miles to the Azores. The captain reckons we'll be there around April 8th, then he's going to stay for a day or two, refuel and stuff like that, and then we go to Lisbon which he says will be another 3 days cruising But he says a lot depends on the weather, obviously it could be earlier or later."

"OK, so you'll be in Lisbon around April 13th?"

"Yeah, I guess so, give or take a day."

Well, I can't finish my job until April 15th, and that means I can't fly out until the 16th and arrive in Lisbon on the 17th.

"How come?"

"Guest airlines don't fly on the 15th."

"Oh."

This meant that Noel would be in Lisbon four days before I got there. It also meant that we only had 6 days to get to Milan. Actually, it meant 5 days, because Guest arrived in Lisbon in the afternoon.

We decided that rather than wait for me in Lisbon and start hitch-hiking from there, Noel would go ahead to Madrid, sign us both up for membership in the Youth Hostel Association and wait for me to fly straight through to Madrid. This would reduce the distance we had to hitch-hike to Milan.

After some calculation on my part, using an old Atlas we borrowed from Mr Berry, we worked it out that it was about 1,000 miles from Madrid to Milan.

If I arrived in Madrid on April 17th it meant we only had 6 days to hitch 1,000 miles.

"That's over a hundred miles a day!" Noel exclaimed.

"More like a hundred and seventy," I said.

"We might not be able to get there in time," he said glumly. "Can you get off work any earlier?"

I went to my boss and looked very forlorn and told him about my problem. He was a really nice guy, Mr Buswell, and right in front of me, he picked up the phone and called the personnel department. In a matter of minutes, it was all arranged. I could finish up on April 12th and fly out on the 13th, which was the evening flight to Santa Maria and on to Madrid via Lisbon. It was all getting very exciting!

"What are we going to take with us?" I asked Noel.

"Huh?"

"You know, clothes, camping gear, stuff like that."

"Oh yeah, well I guess we need to get something like that, just in case we need it."

He hadn't given it very much thought up to now. Fortunately, I had been in the Scouts until I was 16 and I still had some gear. I had a neat little set of a folding knife, fork and spoon that all clipped together in a compact case. I also had a lightweight groundsheet and a canvas rucksack. Noel didn't have any of these things and had to go into Hamilton to look around for any shop that carried camping equipment. He found what he needed in an old Hardware shop called Frith's on Front Street. Going into that store was like going into a dark and mysterious Aladdin's Cave. There were big wooden bins on shelves with every size of screw, nail and bolt you could ever want. All the bins had hand-written labels on them, beautifully written script in ink. There were hinges, brackets, tools and brass fittings for boats. As you walked toward the back of the store it became darker and darker. If Harry Potter had been invented, he would have been right at home there looking for a magic wand or a Cape of Invisibility. They had everything. Noel told me he went in there and after he told them what he was looking for a very old gentleman told him to "Wait right here, young feller" and disappeared into the back of the store.

Noel said that after about half an hour he thought the man had forgotten him and was having his tea break, but then he came out with a rucksack and groundsheet and some other bits and pieces. They were all musty and smelled like they had been in the back of the shop for years. The man had said, "We don't get much demand for this sort of thing." But it was just what he wanted. We were in business.

Now we just had to decide what clothing we should take.

"We should travel light," said Noel.

When Noel said 'light', he meant very light.

"Well, what do you think we should take?"

"A couple of pairs of pants, some change of underwear, a couple of shirts, socks, and a jacket I guess."

"That's pretty light."

"Well, you gotta remember we will be hitch-hiking and carrying it around with us every day."

I had to agree with him. We would be travelling in Spring and Summer through some pretty agreeable countries for that time of year, so it seemed a good idea. I certainly didn't feel like lugging around a lot of heavy clothing all through Europe. Our mothers came up with some suggestions. "Sunlight soap." They both said. "Very sensible. You can always rely on Sunlight and even wash your hair with it." So we had a little canvas bag with Sunlight soap, toothpaste and toothbrush, a safety razor (as they were called in those days) and a comb. That was our toiletries.

We had a trial run at packing our rucksacks a week before Noel left and it all worked out well and looked very professional. Our groundsheets were rolled up and tied to the top of the rucksack, and everything fitted in very nicely. We each had a light-weight waterproof jacket and instead of jeans we had some khaki slacks, so we wouldn't look too scruffy. We didn't want to scare people off from picking us up.

At 8:00 am on Wednesday morning March 30[th,] we all gathered on Albuoys Point to watch the yacht with Noel on it sail off through Two-Rock passage and on to Lisbon.

THE FLIGHT

With Noel on his way as the hard-working deckhand on a millionaire's yacht, all I had to do was get on with my work, buy my flight ticket and look forward to April 13th when I would be flying off to sunny Madrid.

On the afternoon of Wednesday, April 13th my mother gave me a lift to the Bermuda airport. In those days it was a very simple and uncomplicated airport where you just went in, checked your bag in, and an airport worker took it through a door and out to the baggage cart on which it would be towed to the waiting aircraft and placed on board. These days it is a very modern and sophisticated terminal with all the bells and whistles of automatic baggage belts, airport security checks, lounges, shops and boarding gates.

I said goodbye to my mother and the guys who had come to see me off and went inside. I had no idea what to expect. I had no idea what sort of aircraft I was flying on except it was a Super Constellation, whatever that was! All I knew was that my flight was supposed to depart at 6:00 pm and arrive in Madrid at 4:00 pm on Thursday! Even making allowances for time differences between countries it was going to be a very long flight.

I made myself comfortable in the departure area. I sat with the other passengers waiting to board the aircraft. There weren't many, perhaps a dozen of us. We had heard the sound of engines which we assumed was our flight arriving and approximately an hour later there was much activity at the door and then we were invited to "Board the flight, please." As I have said, everything was a lot simpler in those days, and I walked out through the door and straight out through a low wooden gate onto the tarmac and there was my magic carpet to Madrid, crouching like a giant metal eagle waiting to take off. I don't know what I had expected, but I hadn't expected what I saw before me.

Before the day of my flight, I had ridden out to the airport on my bike and looked at a Guest Aerovias aircraft that was on its way to Madrid, just out of curiosity. I wanted to see what I was getting myself into, literally and figuratively. Even so, I was not prepared for the size of it as I walked toward the aircraft to board. Up close it was very impressive, standing very high on its landing gear and having very long wings, while at the rear the three tail fins stood high above the tarmac.

Guest Aerovias de Mexico
Lockheed Super Constellation with long-range wingtip fuel tanks.

When I had checked in for the flight I had asked the agent at the desk how long the flight was going to be.

"The captain is estimating a flight of between eight and a half and nine hours tonight, sir," she had cheerfully answered. When she saw the look of dismay on my face, she hastened to add, "But it could be less if the headwinds aren't as bad as forecast."

I was shown to my seat by the stewardess (nowadays we have to call them 'flight attendants') and settled in. I was toward the rear of the aircraft and it didn't appear to be full. When the door finally closed I still had the row of two seats to myself. I thought that this was a good start because it meant I could stretch out and be comfortable. Of course, I would have preferred a stunningly beautiful young lady next to me, but you can't have everything!

I remember that the inside of the 'Connie' seemed very long and narrow, and there were only two seats abreast. I was sat midway toward the rear on the left-hand side. I was a mixture of excited and slightly scared as the engines started and then revved up as we rolled away from the terminal building.

"Well, here we go," I thought.

To put things into perspective, compared to a modern Airbus A320, the Connie had 4 propeller engines whereas the A320 has two jet engines. The Connie flew at a height of 25,000 feet at 300 mph whereas the A320 cruises at 40,000 feet at 500 mph. The Connie matched the range of 5,000 miles of the A320, which was comforting, even though the flight to Santa Maria was only 2,300 miles. Even so, at 300 mph it was going to take us almost nine hours to get there, allowing for climbing time to our cruising altitude and headwinds. The only real memories of that flight were how noisy it was and how long it was.

As we taxied to the runway for take-off everything seemed to rattle and vibrate. I looked out of my window as we slowly made our way to the far end of the runway and could see St.David's lighthouse as we turned around to line up for our take-off run. I took a death-grip on the arms of my seat.

The engines roared and we began to roll down the runway. Everything was shaking and vibrating and we rolled...and rolled...and rolled. Looking out the window I could see that we were going faster, but we kept on rolling. Looking across the aisle and out the right-hand windows, I could see the bridge across Ferry Reach and thought, "We aren't going to make it!"

Then, as if answering my prayers, we slowly lifted off the ground. As we flew over Ferry Reach and the big Astor Estate I almost panicked because we seemed to be so low that I felt as if I could reach out and touch the small hill just behind it.
Unlike today's jet aircraft that take off at a sharp angle and climb rapidly, we seemed to leave the ground and climb incredibly slowly. I don't know what I expected, but to me, it seemed to take us ages before we reached any sort of respectable height and started to turn away from the north-west runway heading and actually get on our way east toward our destination. As I looked down at the ocean not far below us I began to wonder if we would be flying at this height all the way to Madrid!

The smiling stewardess came down the aisle checking on all the passengers and asked me if I would like anything. I settled for a cup of tea and a blanket. It had started to feel a bit cool and I realised that we must be climbing higher. My memory of the food served on the flight is very limited, but I do know that it was good. I remember it was all very civilised and I was impressed. I guess the fact that the crew knew that they had all the time in the world enabled them to take their time and provide good service to us all. Also, the plane was only about half full and we were well taken care of with only about 50 passengers on board.

Finally, I leaned back in my seat and tried to get some sleep, and that's when the droning started. With my eyes closed, I was aware of every sound that the aircraft made. The four prop engines made a sort of continuous dull droning roar. As I tried to drift off to sleep, I became aware of subtle differences in the sound of the engines. A deeper roar seemed to start on the right-hand side of the aircraft and then it would gradually travel across the cabin and finally seemed to be coming from the left-hand engines.

As this happened, I was aware of a vibration in the cabin that seemed to follow the noise of the engines across the cabin. I realise now that it was a synchronising of the engine noise that I was experiencing. I think that because I was trying to sleep and travelling alone, I was aware of it to a greater degree. Many years later when working for BA I did some research and found a paper on the subject...

"Vibrations from the engines are transmitted through the engine mounts into the wing structure, which in turn excites the whole aircraft body; turbulence from the propellers excites the rear wing which in turn causes vibrations in the rear part of the aircraft. Another important path is through the fuselage in the plane of the propellers; the propeller blades cause very high-pressure fluctuations at the outside of the fuselage which are transmitted into the passenger cabin."

It all made sense because I was toward the rear of the aircraft and perhaps I was more aware of the phenomenon. Whatever the reason, I found it most disconcerting.

Despite the strange vibrations during the long flight, I managed to sleep for the last half of the flight and when I woke up it was the beginning of a bright sunny day. Looking out of my window I could see big white fluffy clouds and below them a huge expanse of a bright blue sea. I looked at my watch and was surprised to see that it was three o'clock in the morning! That couldn't be right, surely? I asked the stewardess and she told me, "Yes sir, it is 3:00 am Bermuda time, but 6:00 am Azores time because of the time difference."
OK. I was going to have to get the hang of this time difference thing.

Just then the sound of the aircraft engines changed and seemed to get quieter and there was a strange sinking feeling in my stomach. I had just started to call the stewardess back to ask what was happening when the captain spoke to us.
" Good morning ladies and gentlemen. We are now starting our descent into Sant Maria and should be on the ground in approximately half an hour."

Surprisingly this was said with a very posh English accent, which I thought was very strange on a Mexican airline.

The aircraft continued to drop lower and lower until I could see small waves breaking on the surface of the sea. Peering out of my small window, I soon saw land. First, it was just a blur in the distance but then it became more detailed and I saw a lot of brown topped with green. It looked like a brown bun topped with green icing. As we dropped lower I could see that the brown was cliffs topped with green vegetation dropping into the sea. We flew on toward the island, always dropping, and we got closer and closer to the cliffs. A little seed of doubt started to grow in my mind. We were getting so low and the cliffs were so high and I was sure we were going to crash into them. I could see the white spray as the waves smashed into the base of the cliffs. Just as I was about to panic, we flew right over the top of the cliffs and the ground appeared beneath us. Immediately the engine noise changed and there was a bump as the wheels hit the runway. We had landed. I breathed a sigh of relief.

We were allowed to get off at Santa Maria, thank goodness. We had been in the aircraft for nine hours and I needed to stretch my legs. The terminal building consisted of one long low white building with lots of windows and what appeared to be a red tin roof. At the far end of it, I saw, as I strolled outside the terminal, a tall square white tower with windows on all sides and a red roof. This was the control tower for the airfield and was the only thing that distinguished the airport from all the other buildings I could see dotted around the countryside behind the airfield. Looking out along the runway I could see that it stretched along the shore and ended at what appeared to be a sudden drop at each end. As far as I know, it is still like this and must be quite exciting for the pilots of large modern aircraft when landing there.

After a short break we were asked to board the flight again and off we went, taxiing out to the runway. It was quite exciting to take off and to look down as we slowly left the runway, seeing the ground suddenly disappearing from beneath us as we flew over the cliffs and the waves of the Atlantic crashing against them far below.

This flight was a lot shorter than the one from Bermuda and we were soon flying over Portugal, seeing the beautiful country below us and then descending toward Lisbon airport above the houses with their red tile roofs. Once again we were allowed into the terminal for a short break. This time it was a much more modern building with shops to tempt the weary tourist into buying souvenirs they didn't need. I resisted the urge to buy Portuguese trinkets that I neither wanted nor could afford.

Soon we were on board again and on the last short flight from Lisbon to Madrid and I was soon walking out of the Madrid arrivals area to find Noel waiting for me. It was the evening of Thursday, April 14th and I had only slept about 4 hours in the last 24 hours.

MADRID-The River Bank Hotel

The next morning I awoke with a start, opened my eyes slowly and looked around. Above me was a lot of dirt and some roots hanging down. I looked down at my body. It was still there, my arms and legs seemed to be attached and I could move them. I had not been eaten by bears, wolves or rats. I had not been killed by a roving band of Spanish gipsies or swept away in a flash flood. All was well.

Let me explain.

When I had walked out of Arrivals and met Noel, I noticed he was carrying what looked like an old potato sack. As we hugged and greeted each other as if we hadn't seen each other for two weeks, which we hadn't, I noticed that the sack gave off the distinct sound of glass clinking on more glass. Bleary-eyed, I walked out of the terminal with him. He was chatting away and telling me of his plans for the night.

"Wait until you see it. It's really cool and will be a great adventure for your first night in Madrid," he was saying.

"So, what? It's like a cool youth hostel or something?" I asked.

"Nah, nothing like that. Wait until you see it. Let's go, it's this way," he said, walking along the pavement and leaving the terminal, taxis and buses behind us.

Assuming we were going out to the main road I trudged behind him with my rucksack slung over my shoulders. Once we got away from the airport and on the main road which was lined with scrubland and open fields, I noticed that he wasn't stopping to try to hitch a ride with any of the vehicles that were passing us.

"Hey, shouldn't we be trying to get a ride?" I asked him.

"It's only a bit further," he replied and just then he turned down a little path that cut away from the road between two fields. "I found this place yesterday when I came out to scout around the airport."

If I hadn't been so exhausted I might have challenged him, but I just followed him down this little pathway that led toward a line of trees about a hundred yards away. There were no houses in sight and it was quite silent as we left the main road behind.

Soon we came to a break in the trees and I realised that we were standing on the bank of a fairly large river, looking down at the riverbed. It must have been the dry season or something because although the banks of the river were about fifty yards apart, there was only a small stream flowing lazily along the centre of the riverbed. The banks of the river were quite high and steep, and it looked like it must have been a large river in the wet season. Noel confidently made his way down a little path that led to the bed of the river and started to walk along to the left, around a bend in the river. I was starting to get worried.

"Where the hell are we going, Noel," I asked him, getting a bit annoyed.

"Here we are!" he said, quite proudly waving his arm toward the river bank. There was a large portion of the bank that had been dug out, presumably by the river when it was at its highest and swiftest. He was grinning at me as if I was supposed to be impressed.

"What?" I said, looking blankly at the cave, for that is what it was, a cave of earth with a large tree above it on the bank, with some of the roots showing through the earth and quite probably holding everything together.

"This is our home for the night!" he announced happily.

"Are you out of your frickin' mind?" I said, getting slightly agitated.

"Hey, what's the matter? I think it's pretty cool. Look!"

He took a few steps up the slope of the river bank and into the dirt cave that extended back into the bank by a few yards. It was quite roomy and he only had to bend over slightly to walk in. Bending over he dug his fingers into the soil at one corner of the cave and grabbed what appeared to be a piece of fabric buried there.

Pulling on the fabric he revealed his groundsheet that he had covered with a thin layer of soil. As he pulled it up I could see his rucksack beneath it.

"I hid this here before I came to meet you. You didn't even suspect it was here, did you? Pretty cool, huh?"

I looked at him, fighting the urge to pick up a stone from the river bed and batter him to death with it.

"You mean for us to sleep here tonight?" I asked, gritting my teeth.

"Yeah man, come on, chill out."

Too tired to argue I rolled out my groundsheet near his and sat next to him. He was sitting down and rummaging in the old sack he had with him when he met me at the airport.

Out came two bottles of beer, and a big bottle of red Spanish wine, a large stick loaf of bread and a huge chunk of cheese. He then lifted out four large candles and proceeded to place them around the walls of the cave and, since it was now starting to get darker as the sun was beginning to set, he lit them and, much against my better judgement, I had to admit it was, actually, pretty cool.

He popped the tops off the bottles of beer and handed me one. Grinning widely, he said "Cheers!" and we clinked bottles together. I looked out of our 5-star accommodation, looked up the river then down, then across at the other bank where there seemed to be lots of trees and shadows and shapes. I thought I could see dark things moving in the shadows.

"Do they have bears in Spain?" I asked.

"I don't think so."

"How about wolves?" I was looking at a shadow in the trees across the river from us that looked an awful lot like a wolf to me.

"I wouldn't think so," he took another drink of his beer. "Not this close to the city, anyhow."

"You mean they do have wolves in Spain?"

"Hey, lighten up man," he said.

Slowly, I relaxed as my sleep-deprived body absorbed the beer and then the wine. We had started on the cheese and bread and it was all pretty good. Suddenly I heard a noise.

"What was that?" I exclaimed. "It sounded like thunder!"

"I didn't hear anything."

"What if there's a thunderstorm in the hills and a flash flood comes down the river?"

Noel looked at me and then started laughing. I grinned and then joined him. It was all crazy and funny and I realised how silly I sounded. We finished the wine and the next thing I knew I had woke up in the morning wrapped in my groundsheet with my head on my rucksack, thankful to be alive and spending the night in a dirt cave with my best friend.

We got up and splashed some river water on our faces, dipped some out in our tin cups and drank it, had some bread and cheese for breakfast, and headed up to the road to hitchhike into Madrid.

MADRID- The Angry Gardener

We got a ride into Madrid quite quickly. A Spanish gentleman had dropped his 'mujer' (wife) off and was returning home. At least that's what I gathered, talking with him in my schoolboy Spanish. Graham Rosser, my Saltus Spanish teacher would have been proud!

He dropped us off near El Retiro Parque and pointed us in the right direction. Noel had already stayed at the youth hostel the night before I arrived and knew the area pretty well.

As we walked along Madrid's picturesque streets, he chatted away.

"Wait until you see this park, El Retiro," he said. "It's amazing."

"Aren't we going to the hostel now?" I asked.

"No. Nobody is allowed until after three o'clock, and you have to leave by ten o'clock in the morning," he said. "but it is really good. It is clean and cheap and you get a shower in the morning and a basic breakfast all for five shillings."

I looked at my watch. It was eleven o'clock.

"So we have four hours to kill?"

"Yeah, but don't worry. This park has so much to see and it is all free."

He was right. We wandered along and came to a stout iron railing fence that seemed to extend out of sight along the street in both directions. Inside was a beautiful paradise of trees, lawns, bushes and flowers. We walked along until we came to an open gate and went in. It was amazing. There was a huge glasshouse with tropical flowers in it that took us an hour to walk through as we admired the wonderful collection of blossoms.

Continuing through the park, we came to a huge statue towering over a fountain and pond. The statue was of a fierce-looking nobleman astride his horse and waving his sword in the air, as they all seem to do, and was magnificent.

Walking away from the statue of Generalissimo Somebody, we found ourselves admiring a beautiful man-made lake with people rowing boats around on it. Everything was like a fairytale dream. There was even a pretty pink playhouse on a small island in the middle of the lake.

"Man, this is something else!" I exclaimed.

Noel grinned proudly as if he were personally responsible for the park's existence.

"I told you so!" he said.

By this time it was almost two o'clock and we were both feeling rather weary. In the middle of a large lawn was a huge tree casting a cool shadow beneath it. It was quite warm and sunny and we both had the same idea. We walked across the grass and sat beneath the tree, leaning up against it after dropping our rucksacks to the ground.

"Aahhh!" we both sighed at the same time.

I leaned back against the tree and closed my eyes, feeling very relaxed and happy. Beside me, I heard something that sounded very much like a cork being pulled from a bottle. Looked at Noel sitting next to me I saw that he had pulled the cork from the wine that had been left over from last night.

He took a swig, went, "Mmmmm!" and licked his lips as he passed the bottle to me.

I raised the bottle to my lips and took a good swallow of the best cheap Spanish plonk that our money could buy. "Mmmmmm-mmmm!" I smiled and handed it back. This was the life, in the middle of a beautiful park in the beautiful

Spanish city of Madrid with good weather and my best pal by my side.

Noel took another drink and passed it to me again. I had another drink and we sat on the grass, leaning against the tree grinning like idiots, surrounded by lovely gardens and flowers. It was so peaceful that we felt as if we had discovered paradise.

Suddenly we were startled by a very loud voice. "Oye! Oigame! Prohibido! Prohibido!" We both sat bolt upright and looked around the park.

Standing on the path some distance from us was a large older man with a weather-lined face and in working clothes with a large straw hat on his head that had seen better days. He seemed to be excited about something because he was waving a rather large machete around in his right hand and glaring at us. Next to him was a wheelbarrow which had some plant cuttings and a rake in it. Being clever detectives, we immediately deduced that he was one of the park's gardeners.

He kept waving the machete and pointing in our direction and shouting, "Prohibido! Prohibido!"

We both stared at him with our mouths open and then looked at each other.

"I wonder what his problem is?" said Noel.

"I don't think he likes us on his grass."

"Maybe we shouldn't be drinking," said Noel, slowly sliding the bottle back into his rucksack while keeping a wary eye on the agitated man who was now hopping from one foot to the other.

Whatever the gardener's problem was, hiding the bottle did not seem to calm him down. He kept waving his machete in the air and shouting at us and then, to our alarm, he started to walk swiftly toward us.

"Uh oh!" said Noel.

"Yeah!" I said.

We both jumped up and grabbed our rucksacks and started to back away from the rapidly approaching and dangerous-looking man. As he got nearer he kept waving his arms and shouting at us. My school Spanish was not good enough to understand the rapid-fire words that were coming out of his mouth, but no translation was necessary to understand that they were not welcoming. We backed away and started to circle behind the tree. We had both decided that a large tree between us and the machete-waving maniac was a very good idea. If he had been wearing an eye-patch and hobbling on one wooden leg with a parrot perched on his shoulder we could not have been more concerned. As we slowly made our way around the tree, I quietly asked Noel, "Which direction is the hostel in?"

He pointed with his chin in the direction of the gardener's wheelbarrow and muttered softly, "It's over that way."

"Okay, when he is on the opposite side of the tree, let's make a run for it," I muttered just as quietly. Slowly we went all the way around the tree until we were back where we started, but now the gardener was on the other side of the tree, still yelling and following us. He had little drops of spittle on his chin which made him look like a madman.

"Run!" yelled Noel, and we both took off, sprinting across the grass and past the wheelbarrow and along the path. Behind us, we could hear the yelling getting more distant. Seeing a gate leading to the street we rushed through it, with some local Madrid pedestrians looking at us with puzzled expressions. Walking quickly along the pavement, we came to a bench and sat down, breathing heavily. Looking at each other, we started laughing.

"Come to sunny Madrid and have a peaceful vacation!" I said.

"Yeah man, I guess he is the local welcome committee, huh?" said Noel in between his laughter. Noel got out his map of Madrid and, pointing said, "The hostel is this way."

MADRID- La Posada

Leaving the park with the mad gardener behind, we strolled along the old streets of Madrid. It was a lovely city and we felt lucky to be there. Soon we turned into a picturesque tree-lined avenue lined with small shops and what appeared to be apartments with balconies above them. I noted with interest that there were quite a few small businesses with signs that read *Taberna* and thought they would be explored by us later in the evening.

Halfway along this avenue was a yellow facade with an open metal grate door with a sign that said: *La Posada.*

"Here we are," said Noel as he turned and stepped inside.

It was dim and cool inside La Posada youth hostel. There was nothing fancy about it. A very attractive young Spanish woman stood behind a low counter at the back of the small entrance hall. The walls were painted different bright colours, red, yellow, blue. It was all very cheerful. On the walls were posters of different views of Madrid.

The young woman smiled as we both said "Hello!".

"Hello. May I help you?" she asked.

Well, we were off to a good start, I thought. I was apprehensive about using my Spanish, even though I knew it was pretty good. At least she spoke good English. It has always amazed me that all the young people of Europe can speak pretty good English, and yet we, the English-speaking travellers of the world, rarely make the effort to learn to speak basic phrases of the counties that we visit.

Well, I thought, I am not going to let the side down.

"Hola, y buenas tardes," I said to her."Como estas usted?"

"Hello, and good afternoon. How are you?"

"Ah, usted habla espanol!" she replied happily.

"Oh, you speak Spanish!"

"Si, pero muy despacio," I replied.

"Yes, but very slowly."

This went on for a few minutes before I introduced us.

"Mi nombre es Juan y mi amigo es Noel,"I said.

"I am John and my friend is Noel"

Noel, who had been standing to one side and trying to follow the conversation by looking back and forth at the two of us as we spoke now beamed happily at the sound of his name.

"Hola!" he cheerfully said to the girl, which took up most of his Spanish vocabulary. I think he also knew *Cerveza* which means beer.

"Mi nombre es Maria," said the girl.

"Hello Maria," I said in English and Noel immediately said, "Hi Maria." He gave her his most brilliant smile.

After this brilliant start, we proceeded to check in to La Posada and the lovely Maria gave us brief instructions before Noel announced that he had stayed the previous night and pretty well knew the ropes. We paid our fee for the night and went along the corridor behind the entrance to look for our room.

"She's a big improvement on Carlos," announced Noel.

"Carlos?" I asked.

"Yes, he was working the night before last night. Nice enough guy and spoke good English also. But Maria is a definite improvement. I wonder if she is free later?" he grinned at me.

"You behave yourself," I warned him. "Besides, they probably have rules about getting friendly with the guests."

"Oh well, you never know," he replied with a laugh.

We went to the room that Maria had allocated to us and found it to be fitted with two double bunk beds. Like the entrance foyer, it was painted in bright colours and seemed to be spotless with a small desk and chair as well as a large cupboard against one wall for storage. We seemed to be the only occupants because the cupboard was empty.

We dumped our rucksacks on the two lower beds with the agreement that if other hikers arrived, I would take the top bunk above him, otherwise, we would take the two bottom bunks. Noel took me down the corridor to show me the large toilet area that had four rooms each with three cubicles, of a toilet, basin and shower. It was all very civilised and clean. On the door to each cubicle was a little sign in various languages that announced "15 minutes per person".

"Does that include shower, shit and shave?" I asked him.

"Yup, you got it. No dawdling allowed. It was enough for me this morning. You'll be surprised. You can always have a second go after the others have used it. It's mainly aimed at people taking too long in the shower and using all the hot water."

Through another door was a fairly large room with four long tables and several chairs at each one. Along one wall was a sink and draining board and a set of shelves with plates and bowls. All very organised and clean.

After we looked around we wandered outside onto the street, smiling to Maria as we went. She really was lovely with her tan skin and long dark black hair cascading over her shoulders.

"I'm in love," said Noel.

"Come on, let's look around," I said, grinning at him.

It was a very colourful and lively street. Each *taberna* seemed to have a slightly different atmosphere. One had some great guitar music coming from deep within, terrific flamenco. There was a guy on the door and when we asked, he indicated it would be the equivalent of five shillings, which was too steep for us.

It was time to eat so we stopped at another tavern and went inside. It was very cool and relaxed with benches and tables and a few locals drinking cerveza and smoking. Everybody seemed to smoke, and they weren't the sort of cigarettes we were used to in Bermuda, which were American. These were obviously self-made, rolled by the smokers themselves and they stank to the high heavens. We sat at a table by the door where there was some fresh air and I ordered *"dos cervezas"* from the waiter. We had a look at the menu and Noel left it up to me because it was all in Spanish, unsurprisingly. As I looked at it I realised that my school Spanish was not up to reading a cafe menu in Madrid. Just as I was about to panic my eyes settled on something that looked familiar: *Patatas bravas con chorizo.*

"Dos patatas bravas con chorizo," I ordered bravely.

Noel looked at me with a mixture of admiration and apprehension.

"Hey, that sounded great. I'm impressed, but what the hell did you just order for me. Did I hear chorizo?"

We were familiar with chorizo in Bermuda. It was a popular spicey sausage and we all had eaten it because all the Portuguese that lived in Bermuda ate it and it had become a popular item throughout the island.

"Yes," I said. "Now we just have to wait and see if I got it right."

He looked at me, not sure what he was getting for his meal.

A few minutes later the waiter, a foul-smelling cigarette dangling from his lips, placed a plate in front of each of us. It was a huge pile of golden cubes of roasted potato mixed with small pieces of spicy chorizo sausage and fried onions sprinkled with some sort of red spicey seasoning. It looked delicious.

"Oh man, this looks good!" exclaimed Noel.

"Graham Rosser would be proud," I said.

"Who?"

"My old Saltus Spanish teacher," I replied, shovelling a forkful of savoury *patatas bravas* into my mouth.

"Yeah, I think he would be," mumbled Noel through an equally large portion that was disappearing down his throat.

It was a great meal washed down with a second cerveza and we wandered back to La Posada, stopping at every tavern door to listen to the music and laughter coming from each of them. It was a Friday night and the people of Madrid were enjoying themselves, and so were we. We went to bed with full stomachs and dreamed of patatas bravas with chorizo.

The next morning, Saturday, April 16th, we woke up after a good night's sleep and, being the only ones in our room, we had a nice hot shower and shaved. In the kitchen area, we had a bowl of cereal and a nice cup of hot tea which was included in the price of the night's stay. It was a really good deal.

"We need to get moving to be in Milan on time," said Noel," but it would be a shame not to explore Madrid a bit more before leaving, so let's stay one more night and then head for Barcelona tomorrow."

We spent all day Saturday wandering around Madrid. It was a very large and ancient city, full of amazing old buildings of unique architecture. There seemed to be statues of somebody famous every time we turned a corner. In one square we came upon a large statue of a skinny-looking old guy on a rather tired-looking horse. He did not have the usual helmet and his head was bald. In his left hand, he had a huge long spear that rested on his stirrup. Behind him was a fat little fellow riding on a donkey that must have been his servant. The bronze statues were amazingly detailed but we couldn't quite understand who they were meant to represent. They did not look very heroic. All the heroic statues usually had the riders waving bloody great swords in the air.

"I wonder who he's supposed to be?" said Noel.

"Some General or other, I guess. They seemed to have lots of them," I replied.

"I dunno, he looks different," said Noel as we stood on the sidewalk across from the statue. We waited until there was a break in the traffic and scooted across to examine the statue and the plaque beneath it.

"Don Quixote!" we both exclaimed at the same time.

Of course it was! Every schoolboy knew of the wonderful story of the old gentleman who wandered the land looking for daring deeds to perform, hoping to rescue damsels in distress and ended up jousting with windmills. He was an inspiration to us all who dreamed of doing great deeds and rescuing fair maidens.

Noel looked up at the misguided knight and wondered out loud, "I wonder if Maria needs rescuing tonight?"

I looked at him and laughed. "The only rescuing that poor girl will need is from you!"

That night we feasted on patatas bravas again. Carlos was on duty. No Maria.

GOODBYE MADRID

It was Sunday morning and we were checking out.

Maria was on duty. "You know today is Easter Sunday?"

We were both surprised. It had never occurred to us.

"We forgot," I replied. "Will it make a difference to our chances of getting a ride? We are hoping to get to Barcelona today."

She looked doubtful. "You would not be able to get that far on any day," she said. "It is six hundred kilometres!"

I did a rapid calculation and realised she was right. Almost four hundred miles. That was a lot of hitch-hiking. We would have to be very lucky.

"It's not possible?" asked Noel.

"Anything is possible," she said with a sweet smile, making Noel melt into the floor, poor guy. What a dreamer.

She went on to explain to us that it was a very big holy day in Spain and many people would be taking family trips or starting vacations and might not want to pick us up.

We decided we had to try, and with sadness in our hearts, especially Noel's, we said farewell to the lovely Maria and started walking to the outskirts of Madrid with our packs on our backs. We soon arrived at the main road leading to Barcelona. It was long and straight, disappearing into the distance through the dry landscape with just a narrow sandy verge that was barely wide enough for us to safely stand on and waggle our thumbs at the oncoming motorists.

It soon became clear that what Maria had told us was quite true. Every vehicle, whether it was an old beat-up pickup truck or a nice new saloon car, was filled with people, mostly families made up of parents and children. The battered pickups had suitcases and bundles of belongings piled high, many with children perched on them and happily enjoying the ride in the back. Many of the children waved at us and some laughed and stuck their tongues out, being cheeky children. After an hour of slowly walking backward and sticking our thumbs out, we stopped and sat on a convenient outcrop of rock by the roadside, despondently waving tired thumbs in the general direction of the oncoming traffic.

"We're never going to get to Barcelona this way," said Noel.

"I know. Every car is full. The whole city must be going to the coast!"

"Well, let's keep walking and hope we get lucky."

A few more cars and trucks swept past us and then we saw a cream-coloured Volkswagen Beetle approaching. Through the windscreen we could see two dark-haired young women in the front. Expecting to be ignored, we half-heartedly waved our thumbs and as it passed we looked back down the road to see if any more cars were coming. Behind us we heard what sounded like a car slowing down. Spinning around we saw that the Beetle had pulled onto the verge.

"Yes!" we both exclaimed and jogged up to the little car. Inside we could see bags and bundles piled in the back seat. We both leaned over to peer into the passenger window at the front which had slowly wound down.

We could scarcely believe our eyes. Sitting in the front seats were two incredibly gorgeous young Spanish women that were as beautiful as La Posada's Maria.

Noel and I looked at each other and grinned. Our luck was back!

Both of us leaned over and looked into the car. They were both in their early twenties and wearing bright colourful dresses. The passenger looked out of the window at us and smiled.

"Barcelona?" I enquired, being the fluent linguist.

"Si, Barcelona!" she replied with a laugh as she climbed out and pulled the back of her seat forward to give access to the rear seat. Noel and I looked in and our hearts sank. It was piled high with all sorts of clutter that our beautiful rescuers were obviously taking on a vacation to the Barcelona area.

"Oh.no senorita!" I said sadly. "No es possible!"

"Si! Si!" she exclaimed. "Es possible!" She and her friend really wanted to give a ride to two handsome chaps such as us.

The driver began to give instructions in Spanish that were much too rapid for me to follow and waved her arms around, and her friend began to move things around in the back in an attempt to make space, but after several minutes it became obvious that they couldn't fit one of us into the small car, never mind both of us. Our beautiful saviour looked at us with such sadness that both of us felt like crying. How we wanted to share a ride to Barcelona with these two lovelies!

Shaking her head, the passenger looked at us and said softly, "No es possible. Lo siento mucho, muchachos." (It's not possible. I am very sorry boys.)

Noel actually tried to climb into the car at this point but ended up squashing one of the packages on the seat, much to the amusement of the girls.

"Come on, Noel. It isn't gonna happen, mate." I said, heartbroken.

The passenger gave us both a hug and kiss on the cheek while murmuring commiserations and then climbed into the car and with a wave from both of them, the Beetle chugged out into the road and slowly disappeared toward Barcelona. Two forlorn travellers from Bermuda stood on the roadside and watched the car dwindle into the distance until we could no longer see it.

"Shit!" said Noel. "One of us could have squeezed into that car, I reckon."

"You would have left me?" I asked him, shocked.

"Damn right I would," he said. "and so would you. You saw those girls, right?"

I looked back at him, thought about the two gorgeous girls, and said, "Yeah, I would have left you, but there really wasn't space."

We both grinned at each other, turned and started walking along the road again, thumbs out, waving in the breeze.

GUADALAJARA

About half an hour later we got a ride. It was an older Spaniard who looked like he could have been a farmer, and his little truck looked older than him. When we waved him down with our thumbs out, he just pulled over to the side of the road and indicated that we should pile into the back of the pickup. There were a few sacks of some rather lumpy and hard vegetables and we leaned up against them and watched as the countryside passed by us. We had only been in the truck about half an hour when our driver slowed down and pulled off the road into a smaller side road and stopped. We both got up and looked ahead over the top of the cab and saw that we were entering a small town. On the side of the road that we had just turned into was a sign that read 'Guadalajara'.

The driver leaned out of his window, pointed down the road we had just entered and said in a loud voice, the way people do when trying to make themselves understood to foreigners, "Mi casa!" and pointed in the direction of the sideroad.

"What's he saying?" asked Noel.

Grabbing my rucksack from the back of the pickup and pushing Noel's in his direction, I said, "He's telling us that this is where he lives."

"This is the end of the line, huh?"

"Yeah, it looks like it."

Turning to the Spanish gentleman I said "Muchas gracias, senor," and waved at him as he put the truck in gear and slowly drove off.

"It doesn't look like much, does it?" said Noel as we looked around.

It did not.

The road we had travelled on from Madrid seemed to pass by the outskirts of the town of Guadalajara. There were a few buildings on the left of the road and in that direction, beyond these buildings, we could see more large buildings and what looked like the spire of a church. To the right of the road there was very little except scrubland with dry soil and a few small bushes.

"We might as well walk along and see if we either get a ride or find somewhere for a bite of lunch," said Noel.

With this as a plan, we started to walk along the road that we knew would eventually take us to Barcelona. We had the same luck as when we left Madrid and were getting quite depressed when we came upon an old stone building with a few cars parked in front and a weather-beaten sign that said 'Taberna la Mancha'. It didn't look like much, but it was time for a cool drink and perhaps lunch so we turned into the dusty car park in front of the tavern and went inside.

It was very old and very dim, and it was filled with the usual foul-smelling smoke from a dozen self-rolled cigarettes that seemed to be hanging from the lips of every person in the room. The room was a long one that stretched the length of the building and there was a full-length bar along the back. There were about eight rough wooden tables with chairs placed along the length of the room, and at the far end away from the door there was a group of older men sat around three of the tables. They were all smoking and talking among themselves as we came in, with a few cups and bottles on each table.

As Noel and I walked through the door we paused to look around and get our bearings. As we did so, the sound of voices from the far tables ceased and every face turned toward us. You could have heard a pin drop. It was as if we were in an old western movie, with us in the role of two strange gunslingers walking through the saloon door. If one of the men had jumped up and yelled, "Git outta town, ya varmints!" I would not have been surprised. Slowly, we walked forward and, as we had done in Madrid, we selected an empty wooden table as near as possible to the door, an open window and fresh air.

We sat back in the rickety chairs and looked around. Except for us, every person in the room was a man in his fifties or sixties, all looking weather-beaten and rather weary. Although they slowly started to talk to each other, we were aware of the less-than-friendly glances that they kept casting our way.

We spoke in subdued voices so as not to be heard and offend the locals.

"Not exactly a friendly bunch, are they?" said Noel with a smile.

"No, I was thinking the same thing."

The waiter came over. He looked quite old and had one leg that had been amputated at the knee and he used a crude crutch to steady himself and walk. As he took our order he seemed to have a permanent scowl. We ordered two beers and he left a menu that seemed to be older than the waiter and as badly worn. I perused it in silence while Noel looked around the room. Being a friendly sort of person, he nodded and smiled in the general direction of the men at the end of the room, but did not receive any response.

"The only thing I recognise is 'jamon'," I told Noel.

"What's that?

"Ham."

"Let's try it. Is it much?"

"You mean in price or how much do you get?"

"Either one."

"It's only a few pesetas, so it's cheap."

We decided to go for the ham, whatever it consisted of, and the waiter came over when I caught his eye. As he approached the table he seemed to be scowling more than ever and when he reached us he abruptly looked at us and said in a very menacing tone, "Italiano?"

His whole attitude was quite threatening and Noel and I looked at each other. The man had presumably figured out, quite easily that we were not Spanish, and was now wanting to know if we were Italians. Very strange.

"No senor, Inglesa." I told him and waited to see what his reaction might be. It seemed to be the correct answer because he suddenly relaxed and almost became friendly as he took my order for the two ham dishes.

"Man, that was really weird," I said to Noel. "It was quite scary. I wonder what he would have done if we were Italians?"

As we watched him go around the end of the bar to put in our order, we heard the waiter say in a loud whisper "Inglesas!" and a murmur went around the group of unfriendly men "Ahhhhh!" and suddenly the room seemed to grow friendlier. A few of the men even nodded and smiled slightly in our direction.

As we waited for our meal, Noel and I looked more closely at the group of men. Half of them seemed to suffer from some sort of disability. Two of them only had one arm. Another, like our waiter, had one leg and a crutch.

"They must be war veterans or something," I said to Noel.

"Yeah, it seems like it," he said.

Just then two more men came through the door, both in their fifties. One had his left shirt pinned back at the elbow. Another one-armed man! The newcomers scowled at us and went past us to join the group at the end of the room. There was a brief conversation and we heard "Inglesa" and then "Ahhh!" as they sat down with their friends.

"I think I know what this is," said Noel. "We did a bit about Spanish history last year in college. These guys are too old to have been wounded in the second world war. I think they must have been in the Spanish Civil War. If I remember rightly it was around 1935 to 1939, and there were some very fierce battles between the two sides and the Italians were responsible for a lot of indiscriminate bombing that resulted in many Spanish casualties."

"Well, that would certainly explain their dislike for Italians," I said.

"Yeah, if we had been Italian, I don't think we would ever have been seen again!" said Noel, grinning.

Our lunch arrived and it was a pleasant surprise, although after the great food we had in Madrid, perhaps it really shouldn't have been. The ham was two large slices of delicious cured ham served with a green salad with sweet ripe cherry tomatoes and tiny cold potatoes boiled in the skin with a spicey coating of olive oil. I think it was the Spanish equivalent to our potato salad. We both became silent as we ate it, and our waiter even came over to make enquiring noises, to which I replied "Delicioso, senor!" and received a smile of approval and said something friendly. As he passed by the old soldiers at the end of the room he muttered something and they all turned and looked at us, smiling and nodding.

"Looks like we're safe for another day, eh?" I said to Noel.

Glad to be having a delicious meal and relieved that we were not going to be mistaken for Italians and murdered, we finished our lunch, paid the waiter and waved to the war vets as we left, receiving waves and smiles in return.

As we walked out to the road, Noel said "Right! Barcelona, here we come!"

We were very lucky. We only had to wait for ten minutes before a man in a battered black car stopped for us. Unfortunately, when we said "Barcelona?" to him, he smiled sadly and said, "No. Zaragoza."

Resigned, we shrugged and piled into the car. Our driver greeted us with a few short words in Spanish and then I ran out of things to say to him in Spanish and we settled into a reasonably comfortable silence with the occasional chat between ourselves as he nodded to some roadside sight and said something which I did not follow. It was all quite lighthearted and the drive was pleasant. It was over a hundred miles from Guadalajara to Zaragoza and by the time we arrived it was time to look for the youth hostel.

ZARAGOZA TO MILAN

Zaragoza might have been a lovely old Spanish city but unfortunately we did not have time to look around it and on Monday morning we left the youth hostel after breakfast and were once more back on the road.

We were very fortunate to get two very good rides and got to Barcelona in the early afternoon. This gave us a little time to explore and we were fortunate enough to see the amazing cathedral that looked like something out of a fantasy movie. We were told by a guide at the door that it was NOT a cathedral but a 'basilica' but we were not able to understand the difference even though he took great pains to tell us. The really amazing thing about it was that it had these towering spires that looked like some sort of tangled web-like structures. All around the cathedral were hundreds of beautiful carvings depicting scenes from the bible. It was a huge building and quite overwhelming, even more so when the guide assured us that it was not yet completed even though it had been started in 1882!

The hostel in Barcelona was almost an exact copy of the one in Madrid, down a side street and in through an iron-gated doorway. There were two bunk beds, the same as in Madrid, but this time there were two German guys there before us. Being typical Germans they had taken the two bottom bunks so Noel and I were stuck with the top bunks. Since the two Germans spoke or pretended to speak very little English and our German was limited to 'Heil Hitler' which we decided not to use for fear of a punch-up, the conversation was not exactly brilliant. We went out and found a nearby tavern and had our favourite patatas bravas. By the time we returned the two Germans appeared to be asleep so we got into our bunks as quietly as possible. I had a suspicion that our two Teutonic friends were not actually asleep because I slipped as I climbed up to my top bunk and made a lot of noise as I fell to the floor but neither of them moved a muscle.

The next morning they were gone before we woke up., which was just as well because they hadn't been all that friendly and would have spoiled my Corn Flakes.

Leaving Barcelona behind we headed up the coast road toward the French border and a town called Perpignan (pronounced perpinyan). It was only about 100 miles up the coast so we hoped to make it in one day easily, and perhaps press on further into France. As it was, Fate had a surprise waiting for us, or me, to be exact.

Not long after we started to hitch-hike a nice middle-aged French couple picked us up. They were returning to Perpignan after visiting friends in Barcelona. They spoke quite good English as well as Spanish so we were able to chat with them quite well. After two hours we waved goodbye to them in Perpignan and started to walk along the road while looking for a ride. It was noon so we stopped at a roadside cafe to have a drink and a light snack.

It was quite a pleasant cafe and we ordered a Coke and then we both saw the same sign at once, with a picture of a delicious dish of ice cream.

"I could murder an ice cream," I said.

"It's certainly warm enough to have one," said Noel, "but I think I will just have a pastry."

This was not unusual. I had always liked ice cream whereas Noel had always been a bread, rolls and pastry kind of guy.

We both ordered and I set about demolishing my large plate of home-made ice cream. Noel was the translator now that we were in France and he was assured by the proprietor that their ice cream was the best in Perpignan. I now have my doubts about this claim.

As soon as we finished we started to walk up the coast road headed for our next destination of Nimes which was another 100 miles along the southern coast of France. We hadn't gone very far before my stomach started to feel very strange and I said to Noel, "I think I have a problem."

Noel looked at me. "What's wrong?"

"Oh man, I don't think that home-made ice cream agreed with me. I have to find a toilet!"

"You're joking, right?"

"No mate, I'm deadly serious."

At this point, I had turned around and was walking briskly back to some public toilets that we had just passed on the roadside. After I emerged, Noel looked at me and said, "Man, you're as white as a ghost!"

"That's about how I feel, too" I said.

"You up to hitch-hiking?"

"No mate. No way. I wouldn't feel comfortable getting in a car right now."

As soon as I said this, my stomach started its rebellion again and I lurched toward the toilets with Noel watching me with mounting concern.

When I emerged about fifteen minutes later he announced that he had questioned a passing local about hostels in the area. He had obviously realised that I was in a bad way. There was a hostel nearby and we slowly made our way there with me hoping it was open because my stomach was grumbling again. I don't know what was in the 'best ice cream in Perpignan' but it certainly didn't agree with me!

We had not planned to spend the night in Perpignan but alas, we did. Noel left me in the room while he went out to buy some cheese and a large baguette, or stick loaf. I slowly recovered enough to eat a little and the next morning I was able to get up and eat some breakfast and we set out on the road to Milan once more. It was now April 20th and we only had three more days to get to Milan which was almost 500 miles away!

We walked along the road and soon came out into the countryside. It was a lovely day but nobody was interested in giving us a ride.

We trudged along the southern French countryside, doing our best to enjoy the scenery and the lovely weather and not worrying too much about the fact that we had only three days to travel five hundred miles! Every time we heard a car approaching we spun around and put out our arms and waggled our thumbs. Writing about it now reminds me of my time in Africa many years later.

I had been sent to work in Nairobi airport in Kenya by British Airways for a month's experience. Because of the airport's height above sea level and the high heat during the daytime, all the flights came through Kenya at night when it was cooler. Height and heat make for less dense air which makes it difficult for large heavily–laden aircraft to get airborne. As a result, I had lots of time to myself during the days to go exploring in Africa. On the day that I remember, I had taken the company car and headed for a place called Amboseli, which was a national park and, according to the locals, a good place from which to view Mount Kilimanjaro. So, off I went on my own, into wildest Africa. Idiot.

The roads in Kenya were basically red dirt roads, fairly smooth but very dusty. It was quite warm and as I travelled through thousands of square miles of flat plains and thorn trees, marvelling at the sight of zebra, giraffes and warthogs, I found myself approaching what appeared to be about a dozen mini dust tornados. Keeping an eye on the closest and largest one, I timed it so that I let it cross the road about a hundred yards in front of me while avoiding the smaller tornados that were all around me. Having survived that ordeal I reached the lodge that had been my destination, had a light lunch, took some photographs of Mount Kilimanjaro as it peeked through the clouds that always seemed to cover it, and jumped in my car and headed back for Nairobi.

The road was dead straight and seemed to disappear into the horizon between the thorn trees. Up ahead I could see a smudge of bright orange. Puzzled, I kept going at about forty miles an hour but kept an eye on the object. As I got closer it slowly became clearer. It was a seven-foot-tall Masai warrior. He was balanced on one leg and using his rather nasty looking spear to keep his balance as he waggled his hand at me in what I took to be a hitch-hiking gesture.

In a few short seconds the following thoughts flashed through my brain:
Where had he come from? There didn't seem to any sign of a village or roads.
Was he friendly? Were they cannibals?
Should I give him a ride? Would he murder me and take my car?

 I was in the middle of nowhere and alone in this man's country.

I know now that I could have given him a ride and that the Masai are friendly people. I put my foot down and looking straight ahead I swept past him. In my rear-view mirror I could see him being enveloped in the red cloud of dust that the car was trailing behind it and he was looking at me impassively, simply accepting that the white man had decided not to give him a ride. I have never forgotten that instance.

Meanwhile, back to April 20th 1960, Noel and I were having the same luck as the Masai warrior. Nobody was stopping for us.

We kept walking and trying to get a ride to no avail. It was pleasant countryside and the weather was good. Soon we came to a small village beside the road. There were large farm fields on the left with the village across from them on the right. We sat on a low wall and rested for a few minutes, looking around us.

"What the hell...?" exclaimed Noel.

"What's wrong?" I said, looking at him. He was staring back down the road we had just walked along.

"We have competition," he said solemnly.

I looked down the road and sure enough, I saw two figures standing on the roadside trying to hitch-hike.

"Oh crap," I muttered.

Whoever they were, if any friendly driver came along, our two competitors would obviously get a ride before us, making life very difficult. We decided to sit on the wall and let them catch up to us. It's always nice to size up your competition.

We watched as they slowly made their way toward us. Although they were dressed in clothing similar to ours, it soon became clear that they were very different from us, because they had longer hair and their bodies looked a lot more appealing. They were two young women about the same age as us.

Noel turned to me and smiled. "Well, well, well," he said.

"OK, they're nice looking, but they're still competition," I warned him.

As they reached us the two girls smiled broadly and greeted us with a chorus of "Bonjour!" and "Ca va?". One was a blonde and the other brunette. They had backpacks, the same as us, were very attractive and looked very fit.

Noel was our French linguist and he ascertained that they had come from Barcelona and were heading to Nimes, the same as us, but from there they were branching off for Paris, where they both lived. We were faced with a difficult decision. We could walk with them and keep trying to get a ride or we could be gentlemen and walk ahead of them so they could get the first ride. It would be nice to have the company for a change but we all knew it would be much harder for four hikers to get a ride. We decided to stay together for a while but after an hour we gave up and they let us get ahead of them. After another hour it was getting late and we were out in the countryside with very little hope of a ride but worse still, no obvious place to stay for the night. We stopped in front of a roadside cafe and let them catch us up.

We were all quite depressed by this time and the conversation was sombre, with Noel doing the talking and me trying to follow it. While they talked, I looked around us. A hundred yards along the road was a large field with some very large haystacks scattered around in the middle. They must have been fifteen feet high and twenty feet wide.

"Man, they build big haystacks out here," I observed.

Three pairs of eyes turned to stare at me.

"What's that?" said Noel.

"Oh, nothing," I replied. "I was just saying how big those haystacks are."

He was grinning like a hound dog. "John, you're a genius!"

"I am?"

He had turned to the girls and animatedly talking in French and pointing toward the haystacks and then the cafe. At first, the girls seemed dubious but first one and then the other started to smile and look at each other and with a few very Gallic shrugs said " Pourquoi pas?" and giggled.

"What's happening?"

"We're spending the night in Haystack Hotel, mate" grinned Noel.

The four of us went into the cafe and bought some bread, cheese and wine and thus laden we walked up the road until we were out of sight of the village and then we slipped through the hedgerow and jogged out to the middle haystack and went around to the side away from the road.

Up close, the haystacks were huge. We started to work our way inside them, pushing and packing the straw until we had made a fairly large cave that went about six feet into the haystack and was about four feet high, enough for us to sit up in or spread out to sleep. We soon had our groundsheets spread out with the food and wine laid out. It had all the ingredients of a delightful picnic. Corks were pulled and soon I was talking fluent French, much to the amusement of Noel and the girls, who only understood every other word, probably because my French was terrible.

There was much hilarity and good fun. Sitting in our haystack hotel room with a view we watched the sunset and finished off the wine and with some sloppy kisses and "Bon nuit" we drifted off to sleep. It was 1960 and nothing else happened, dear reader. Perhaps now, in the twenty-first century, it might have been different, but it didn't matter. It was a great night.

The next morning we pulled straw out of our hair, exchanged addresses and returned to the road after hiding all signs of our presence at our hotel.

Back on the road, we went on ahead and within half an hour a car passed us with the two girls in it waving like mad as they left us behind. We didn't mind. Our turn was next. Soon we had good luck also, and a French gentleman picked us up and took us right into Nimes which we reached before noon.

It was now Thursday, April 21st, and we still had three hundred miles to cover in a maximum of two days.

MILAN

We got lucky. A large commercial goods lorry driver gave us a ride to Marseille where we stayed the night in a truckers' hostel where he was staying and the next morning he was going to Genoa and offered to take us, I think because we had helped him unload his cartons of toilet paper in Marseille. It was a great ride along the coast, especially because we were sitting up high in his cab and got great views of the coast. In Genoa we helped him unload some more cartons and then he told us where to go for the best chance of getting a ride to Milan and sure enough, we were in Milan on the afternoon of the 22nd.

We had made it. The next two days were occupied with learning all the details of the apartment of Mr.Berry's friends. Noel had met them before and they seemed to be very nice people. They must have been, to hand over their lovely Milan apartment to two young hitch-hikers! We were surprised when, just before they left, the gentleman handed over a large amount of cash that Mr.Berry had asked him to give to Noel.

We had brought as much cash as we had managed to get together before we left Bermuda, in the form of American Express traveller cheques that we had bought from the bank in Bermuda, but we guessed that Noel's mother had pestered and worried his father and persuaded him to make sure Noel had enough money for the rest of his trip. It was a welcome surprise because we did not have very much money between the two of us. The first thing we did the next day was go to the bank in Milan and buy more traveller cheques.

The next thing we did was head for the centre of Milan to visit the Duomo of Milan, Il Duomo, the world-famous cathedral. To be honest, I did not know much about it but Noel's parents had told us that it was one of the things we had to see "if you do nothing else on this trip!".

We were glad we did. Il Duomo is one of the most impressive sights I have ever seen to this day, sixty years later. The basilica in Barcelona was pretty impressive but this cathedral was overwhelming. As you approached it you started to realise how insignificant we all are in the scale of things. Il Duomo was enormous with spires reaching up into the sky that seemed to go on forever. At the centre was the main door that has many sculpted bronze panels on it. The massive doors themselves were closed and must have been at least thirty feet high. We lingered at the rear of a tour and heard the guide tell his group that the doors were rarely opened except on special holy days. The detail of the panels was amazing and told the story of Mary and Jesus. Many features on the lower panels had been rubbed smooth and shiny by the many worshippers. The one that caught my eye was where Jesus was holding Mary's hand and both hands were rubbed smooth by the millions of hands that had touched them. It gave me a funny feeling to see this. As we listened the guide told his group that the cathedral was capable of holding 40,000 worshippers. Noel and I looked at each and laughed. "That's damn near the entire population of Bermuda!"

At that time in 1960, the population of Bermuda was just over 44,000. These days it is approximately 60,000 people. We spent the entire day wandering around the outside of the cathedral, captivated by the size of the structure and the ornate artwork that adorned it. Inside it was like a massive cavern with more statues and artwork. Truly a memorable visit.

The next day, Noel had a surprise for me. I had heard of the famous painting of the Last Supper but had not realised it was actually in Milan. "We really have to go look at it," he insisted.

Off we went to the church of Santa Maria delle Grazie. As we arrived at the address we thought it must be a mistake. After the majesty of Il Duomo, the neighbourhood was remarkably average. The building was a modest one in a side street. We went into the entrance hall and were met by a very friendly Italian priest who was obviously the curator. He was a short stocky man with a perpetual smile that grew wider when we both dropped some paper money into the donation box.

He guided us into a large room that looked like it had been repaired or re-built fairly recently and it was empty with a concrete floor that had been swept clean but otherwise very humble. As we entered we looked at the far wall and there it was: The Last Supper.

I have never been terribly religious, but the sight of this masterpiece made me quite emotional. Leonardo de Vinci had captured the moment of the twelve apostles sitting at the table with Jesus perfectly. The expressions on all their faces told the story and made you understand it even if you had never heard it. The fact that the painting was faded and slightly damaged did not detract from it but made it all the more powerful. I understand that, sixty years later, it has been restored significantly, but I do not believe it can have a greater impact on a person than it had on me that day.

More was to come. As we left the room in which the painting was situated, the curator took us into an adjoining room which was bare except for dozens of large black and white photographs. It took a few minutes for me to realise what I was seeing.

In his emotional Italian and broken English, the curator explained as he pointed to each photograph. One showed a neighbourhood of rubble with only a few chimneys and walls still standing.

"Big bomb," he said.

Shaking our heads, we both said, "Bloody Germans, eh?"

"No signori," he shook his head sadly and looked at us. "Inglese e Americano."

Noel and I looked at each other, aghast. It was true. As we looked around and found the notes on the photos that were written in English we learned that the Allies had bombed Italy in 1943 and the photos that we were looking at showed the results of a large bomb dropped near the church.

As we looked at more and more photographs we found ourselves becoming shocked at the devastation we were seeing. It seemed the whole neighbourhood had been flattened. The last few pictures were unbelievable. Amid the rubble was one wall still standing. Taken from the back, it looked like every other lone wall in the area that was still standing, but then the next picture was taken from the other side, the side that had the Last Supper painted on it. It was still standing amid the total destruction around it. How can you explain something like that? We looked at each other, then at the curator, who had a huge smile on his face.

Nodding his head slowly he said, "Un miracolo!"

We couldn't argue with that.

COMO

The owners of the apartment were away for a month so we had lots of time on our hands to look around Milan before we set off on the rest of our journey around Europe. We spent the next few days sightseeing around Milan. It was a beautiful city but we soon became restless.

Noel was reading the local newspaper, "I was thinking."

Uh-oh.

"What about?" I asked.

"We weren't very successful with our hitch-hiking. I think we need our own transport."

"I don't think we can afford a car!"

"Well, we have quite a bit of cash now, after that bundle we got from my dad."

"What's your point?"

"There's an advert here for a Vespa scooter," he said, waving the paper at me.

"Ah."

"How much is fifty thousand lira?"

I almost choked, but then I realised that it wasn't that much because the Italian lira was something like two thousand to the British pound.

"I think it's about twenty-five pounds."

He beamed at me, "Then it's no problem. We should go have a look at this scooter."

We got out the map of Milan and it looked like the address was just around the corner, so we set out to find it. It turned out to be around five corners, but we found the small bike garage without much difficulty. The Italian in charge was a friendly fellow who rolled out the Vespa for us to look at. It was bright red and spotless. In broken English, he told us that it was his sister-in-law's and he had serviced it for her, but she was now pregnant and didn't want to be riding it any longer.

With the garage owner's permission, Noel jumped on the scooter and rode it around the block and came back looking very pleased. I took it for a test and agreed with him, it was a good buy. Fifteen minutes later and twenty-five pounds lighter we had a bright red scooter and a bill of sale. We had our very own transport.

We spent the rest of the day just cruising around Milan like locals, dodging the traffic and blowing our horn at the pretty girls sitting at the roadside cafe tables.

Back at the apartment, we discussed our next move. We had a few brochures that were lying around the apartment and both of us like the looks of a place called Como. It looked like a very pretty town at the southern end of a long narrow lake. The pictures looked amazing, as do all travel brochure pictures.

Looking at the map it looked like it was only about thirty miles north of Milan so it wouldn't take us very long to get there.

"We'll leave early in the morning, cruise up there, look around, have lunch and cruise back. No problem."

The next morning we had filled up with petrol and set off at 8:00 am, heading north. It was easy going at this time of day because most of the traffic was entering Milan as we were leaving it. Soon we were out in the countryside and passing lovely fields of what looked like olive groves and an occasional small field of grapevines. It was absolutely beautiful, made even more so by the fact that we were two free spirits on our own bike without a care in the world.

There is something about Italian architecture that appealed to me then and still does to this day. The terracotta tiles on the roofs of cream or yellow houses were so attractive. Now and then we would see a larger house with a tower at one end as if the owner wanted everybody to see how rich and successful they were.

Following the road signs we finally entered Como, and it did not disappoint. Nestled at the foot of Lake Como with hills all around it, the town was like a perfect postcard scene. There was a lovely area on the shore of the lake with lots of cafes and restaurants offering alfresco dining and, after a wander around the shore-front, we grabbed a waterside table and ordered a couple of beers and sat back, fully pleased with ourselves.

Well, what do you think?" asked Noel. "Isn't this perfect?"

I heartily agreed with him. Looking around at the hills, the water and the picturesque houses, it was hard to imagine a more pleasant location to have lunch. A large number of boats bobbed gently in the lake and the rigging from the sailboats gave an almost musical background as it rattled against the masts.

"Saluti!" I raised my glass to Noel.

He returned the gesture. "Cheers, mate."

I do not remember what we had for lunch, but I believe it was a nice seafood plate that cost almost nothing and was delicious. The Italians can cook as well as build nice houses. We sat for a while, completely relaxed and satisfied and reluctantly got on our new red Vespa and headed back to Milan. It had been a throughly wonderful test of our bike and our ability to find our way around the country.

With our scooter, we began to explore the area around Milan, and there was so much to see. Italy has fantastic little towns and villages perched on hilltops, all with such attractive architecture. Also within a short ride from Milan were some beautiful larger towns and cities so we decided to visit a few. On top of the list was Verona, which was the furthest away. We got up one morning and left before the rush hour traffic and reached Lake Garda at 10.00 am. Lake Garda was on the way to Verona and is similar to Lake Como with its surrounding hills and villas perched on the hillsides.

We had a coffee at Garda and then pushed on to Verona, which was the town featured in Shakespeare's Romeo & Juliet, and, naturally, we joined a tour without paying so we could follow them to the little square outside the house of Juliet and see the balcony where she stood and said something like, "Hey Romeo, where is you,boy?" although I think William said it a bit more eloquently.

On quite a few days we simply got on the Vespa and rode, following roads at random and were never disappointed. It seemed as if we couldn't go wrong. One of our trips took us to a town called Bergamo and it was beautiful also.

On another day we rode the short distance to the town of Monza where the Formula One Grand Prix was held. We weren't huge followers of Formula One like lots of other people, but we had been thrilled by the story of Stirling Moss a few years earlier and how, at Monza, he had run out of gas but still managed to win the race after refuelling! There were no races on but we were allowed to go in and look around, and there was a Ferrari museum that was free. We went in and admired the gleaming, sleek sports cars and joked about how all the Ferraris seemed to be red.
"A bit like Henry Ford telling the American people they could have any colour Model T as long as it was black" we joked.

Before we knew it, the days had flown by and the apartment's owners were due back. We spent two days staying home and tidying the place up as much as two young single guys knew how, and the owners seemed satisfied when they came back.

THE ALPS

We were looking at the map of Italy that we had found in the apartment, getting advice from Mr Berry's friends. There was also a large map of Europe and we had both of them on the dining table, trying to figure out how we were going to get from Milan to Paris on our little red scooter.

"Wow!" I said, looking at the area to the north-west of Milan, between us and Paris.

"Yeah," so Noel, rather glumly.

The north of Italy is surrounded by a small obstacle called the Alps. They curved all around the top of Italy and formed a natural barrier between Italy and France, Switzerland and Lichenstein. Long before political boundaries were agreed between countries, natural features such as rivers and mountains had become the dividing line between territories claimed by different tribes of people. It made simple sense to accept that your territory ended where the river or mountains made it obvious where to draw the line and let the other tribe claim the area on the other side.

As we looked at the map, the idea that we could just head straight for Paris from Milan became ridiculous.

"If we go up here," I said, pointing to the north-west, "we seem to come up against these little hills called Mont Blanc and The Matterhorn."

Noel laughed, "Yeah, those hills are about thirteen thousand feet high!"

"So we have to go around them?"

"Yes, back through Turin and then turn up north toward Lyon. It looks like there's a pass that goes through here," he said, running his finger along the road past Turin and into France over the Alps.

We did not appreciate the fact, that as you climb, the temperature drops by as much as four degrees Fahrenheit for every thousand feet that you climb. We were about to find out.

The next day, May 27th, we set off bright and early. It was a nice sunny day and we had on our Bermuda clothing of slacks, a shirt and windbreaker. We rode out of Milan in high spirits and ready for anything. I was driving and it felt good, the wind in my face, Noel sitting behind me and commenting on interesting sights. We made good time and expected to get to Lyon by the evening. It was only 270 miles.

Soon, the crowded streets of Milan were behind us and the stunning countryside was being passed as I kept to a conservative fifty miles per hour. I always got the impression that, whenever I drove, Noel was squirming around on the seat behind me, wanting me to go faster. He loved speed.

After an hour we changed over and sure enough, we went much faster as Noel open up the throttle. It wasn't long before we could see the first hills in the near distance, and before we got into them, we decided to stop for a brief lunch break. We found a roadside tavern and had some light refreshments and then proceeded on our way again, me driving. The terrain started to rise ahead of us and soon we started to climb slowly.

Before long we were in the foothills of the Alps and our speed dropped as the road became steeper, more winding and narrow. We realised that our optimistic hope of driving to Lyon over the Alps in one day was highly unrealistic. Before long the air started to feel colder and we had to stop to get a sweater out of our backpacks.

We were still climbing as the day started to become darker and even colder. Before long each one of us was only able to drive at the front for twenty minutes at a time with the other person sheltering behind out of the cold wind.

Sitting behind the driver, it was possible to hunch down and stick your hands in your pockets and try to stay reasonably warm, but the person driving the scooter was exposed to the cold wind and, having no gloves, our hands would become unbearably cold after fifteen minutes or so.

It was almost eleven o'clock when we both decided, we had to stop and find shelter. But where? We were on a winding mountain road with sheer rock faces on one side and an almost vertical drop on the other side and only an occasional house every few miles. Just as we were getting desperate, we realised that the road had started to descend. In the dim light with mountains all around us, we could not see far ahead to know what the terrain was like and could only hope to find a village or inn.

Just then, visible ahead, we could make out what appeared to be a widening of the surrounding countryside and the steep walls of rock melted away and there was a valley ahead. We were on a gradual downward slope in the road and there were only a few lights on in what appeared to be a small village. Suddenly Noel, who was in the front seat, put on the brakes and brought us to a stop.

"What is it?" I asked.

"Look," he said, pointing ahead with his chin.

It was a large truck pulled off to the side of the road. It was parked in front of a small cottage that was set back from the road.

"What have you got in mind?" I asked.

By this time he had pulled the scooter off to the side of the road, a few yards behind the truck, and turned the engine off.

"Let's go check it out," he whispered.

We both got off the bike and Noel pulled it onto the stand. Looking like two furtive criminals we crept up to the truck, and Noel tried the driver's door. It opened.

"We can shelter inside here," he whispered.

"It won't be much warmer inside the truck," I complained.

"Better than on that damned bike," he replied.

Trying to make as little noise as possible we both climbed up into the cab of the truck. We had our backpacks with us and pulled on as many extra pieces of clothing that we could find and wrapped our groundsheets around us. It made a slight difference and the temperature in the cab slowly became bearable as we settled down. We drifted off to sleep, as comfortable as possible.

Several hours later we both slowly woke up as daylight started to fill the valley and we were startled when another lorry and then a postal van drove past us. We looked at each other. It was time to leave. I sat up, as did Noel, and he started to unwrap his groundsheet from around his body and packed it away into his rucksack. I just sat there with a feeling of panic overwhelming me. I couldn't move my right hand. It was bent at right angles at the wrist and frozen in place where I had draped it over the steering wheel during the night. As hard as I tried, I couldn't move it.

"Let's get out of here before the owner comes," said Noel. He had not noticed my predicament.

"I can't," I moaned.

"What do you mean, you can't?"

"I can't move my hand!" I held it up to let him see. It looked grotesque. The hand was locked in position, bent at the wrist. No matter how much I tried to move it, I couldn't.

I tried to straighten it with my left hand but it refused to move without hurting. It had been bent over the steering wheel all night and now it was frozen in place. I was starting to panic, thinking I was going to lose my hand or something equally terrible.

"Slide out of the cab, we can work on your hand after we get out!" said Noel as he scooted over toward me. We didn't want to open the other door which was facing the cottage.

Awkwardly I slid over and opened the door on my side with my left hand and then stumbled out of the cab, pulling my bits and pieces with me. My backpack fell to the ground. Noel followed me out of the truck and silently closed the door. We crept back to the bike and Noel helped me pack my clothing and groundsheet away and then we jumped on the bike and started to roll it quietly downhill with me sitting on the back and holding on to him with my one good hand. I managed to jam my paralysed hand into his jacket pocket from behind. As we gained speed Noel let out the clutch and the bike started, and we began our descent into the rest of the valley. Soon, I felt the feeling creeping back into my hand and when I pulled it out I found that I could move it slightly. Before long I could move it almost normally and a feeling of great relief crept over me, I wasn't going to lose my hand after all!

We set off as the dawn light was starting to fill the valley. We descended from the high pass that had brought us through the mountains and as usual Noel was driving like a lunatic.

I was holding onto him with one hand while my right hand was thrust into his jacket pocket from behind in the hopes that any warmth it found there might help it to recover. Portions of the road descended quite steeply and as we swept down a steep little cutting between two cliff faces the road suddenly opened out in front of us and I had to keep myself from screaming with terror as, over Noel's shoulder, I saw a small village square straight ahead of us, surrounded by two-story stone houses and with an ancient fountain in the centre of it, dead ahead of us. It all happened so quickly that it looked as if we must crash headlong into the rough stone fountain and thereby end our adventures in Europe in one spectacular crash.

I heard a startled expletive from Noel and then he wrenched the handlebars around in a frantic swerve and, as we missed the fountain by millimetres, he dragged them back in the opposite direction and as lights came on all around us in the village houses we disappeared from the village square and along the road with loud shouts of laughter and relief from both of us, probably more so from me. To this day I do not know how we escaped death in that small Alpine village, but I do know that Noel never let up on the throttle and we were so synchronised when we rode that as he had thrown the bike and himself to the right and left, I had copied him.

Perhaps that is what saved us, we were the perfect team.

We rode for an hour with Noel in front and then I told him that I thought I could control the bike by then because the right hand that controlled the throttle had thawed out sufficiently. We changed over and I could see that Noel was almost frozen from being in front for so long.

"Sorry mate," I mumbled.

With me at the controls, we set off at a more sedate pace, with Noel muttering in my ear from behind. At some point, we passed through a border checkpoint manned by some sleepy French border guards. In 1960 the UK was not a member of the Common Market as it was called then, and we were both using British passports (even though I had an American passport as well) but it did not seem to worry the guards and they waved us through after a casual inspection. As far as they were concerned we were just two young lads on a scooter mad enough to cross the Alps in what must have appeared, to them, summer clothing!

Soon we were in the warmer countryside of France and we entered Lyon in time for lunch in a quaint inn. After lunch, we took a room at the same inn and had a hot bath and fell into bed exhausted.

LYON

We had intended to push on to Paris without stopping in Lyon, but when we mentioned this to our host at the inn during a breakfast of delicious croissants with homemade jams he was so insistent that we stay and see his city that we decided to stay with him another night.

As we strolled around the city we were glad that we had stayed the extra day. We knew nothing about Lyon but soon realised it was absolutely beautiful, situated as it was on the river Rhone, with splendid buildings and many ornate churches. The centre of Lyon was full of elaborate old buildings and there were the usual cafes with seating outside, many overlooking the river or the lovely parks that seemed to be around every corner.

Ou host had told us about an old Roman outdoor theatre and we caught a bus that dropped us off at the gates. We paid a small entry fee and were astounded at the semi-circle stone seats built into the side of the hill with the stage in the centre below. According to a guide who was on duty, it could seat 10,000 people which was a quarter of the population of Bermuda.

I hate to admit it but, as lovely and interesting as the city was, the thing that I actually remember was the great meal out host prepared for us that night.

We were the only guests and I think he enjoyed himself preparing what he described as a true Lyon feast. It consisted of the tastiest sausages I had ever had, with potatoes that had been sliced with the skins left on and fried with herbs and onions. We ate until I thought I would burst, washing it down with a bottle of great red wine that he insisted was free because we had stayed to see his city at his insistence.

We slept the sleep of well-fed men that night and set off for Paris the next morning after another breakfast of fresh croissants and coffee.

THE CAFE

It was 250 miles to Paris from Lyon and we made our way through the countryside at a sedate pace. Even Noel seemed to be relaxed after our satisfying sojourn in Lyon and he seemed happy to let me drive most of the way.

As we rode along the tree-lined roads that seemed to stretch straight as an arrow for miles, it dawned on us that these were old Roman roads. The Romans had built many roads throughout Europe and France seemed to have their fair share of them. It was very pleasant driving on them with the sun flashing through the trees at intervals as we enjoyed the scenery.

We were learning a lot about how the different countries of Europe, although part of one huge landmass, all had their own unique culture and architecture. We had now visited Spain, Italy and France, and the cities of Madrid, Milan and Lyon, while sharing similarities all had their own style. The people also had, over the ages, developed different personalities and habits. Each country had welcomed us with warmth and friendship and we had discovered how all the people acted and dressed in a way unique to the country. It was quite fascinating.

At the time of our trip, it was still obvious that it was only fifteen years since the end of that terrible war, and especially in Italy and France, we had seen evidence of the damage that had been inflicted on the country. Bermuda had been isolated in the middle of the Atlantic and had only been affected in the smallest of ways compared to the Europeans. We could only wonder at how it had affected the people, but so far we had met only warmth and good humour.

As I steered the Vespa along a pleasant country road I spotted what appeared to be a quaint roadside inn up ahead. I slowed down and pulled to a stop in front of it. Although it had an atmosphere of slight decay, it had a certain charming appeal and we decided to give it a try. After all, we only wanted a drink and a light snack.

It was a nice warm day and there was a grapevine covering a seating area in front of the inn with fresh green leaves allowing the sunlight to filter through onto the wooden tables below. Parking the scooter we went to sit at one of the tables and looked around. We were the only patrons and it was all very peaceful. After a few minutes, Noel got up and went up to the door and looked inside. I could see from where I was sitting that it was quite dim inside the inn.

As Noel tried to get some service, I had time to look at the inn more closely. It had seen better days and although it appeared to be kept clean and tidy, I suspected that it was not a thriving establishment. From my seat at the table, I heard Noel speaking in French to somebody who was out of sight and then he came back to the table and sat down. He was followed immediately by an older French lady dressed in black. She could have been anywhere between fifty and seventy and her face was the face of a woman who had seen trial and tribulation. She smiled and made an effort to be pleasant while taking our order but I could see her taking stock of us, almost examining us, our clothing, looking over at the Vespa and listening to us as we talked to each other in English. She and Noel exchanged words and he laughed and shook his head and then she walked back inside the inn.

"What was that all about?" I asked him.

"She seems to think we are rich Mafia guys from Italy," he replied, grinning.

"Why the hell would she think that?" I asked.

"Because I told her we had come from Milan and were going to Paris. I think in her mind that is what members of the Mafia do," he laughed.

"So what did you order for us?"

"I asked for two beers as well as some pastries and cheese," he replied.

Just then a young woman came out of the inn carrying a tray with two bottles of beer and two glasses. She was closely followed by the old lady who hovered nearby and watched the girl serving us. She seemed to be taking a keen interest in the proceedings.

The girl who served us could only have been in her late teens, perhaps 17 or maybe 18, no more. She was dressed in a rather dull grey dress that did nothing for her and her dark hair hung loosely to her shoulders. As I looked at her our eyes met and she gave me a shy smile and then she quickly became serious again, almost as if she was afraid of being caught at doing something she shouldn't. In that split second, I had seen that she was quite pretty with startling blue eyes.

The girl turned away from our table and hurried back into the inn, but I glanced up to see the old woman watching me intently and then she hurried toward the door, talking French rapidly in a loud rasping voice, obviously to the young woman. I looked at Noel.

"What the hell?"

He shrugged, "Who knows?"

We sipped our beer and talked casually about what we might do next, especially in Paris, and then the girl reappeared with our pastries and some cheese on a tray. Surprisingly, she had managed to change in the short time and was now wearing a very attractive flower print dress that fitted her better and was cut low in the front so her nice figure was more obvious. Her hair had been tied back with a flower tucked behind her ear. The change was so obvious that we both stared at her.

The old lady had come outside again and was standing nearby as the girl served us. She seemed to be watching us, and especially me, more than the girl.

I smiled at the girl as she placed the tray on our table and pointed at her dress and said, "Tres jolie," which I hoped meant 'Very pretty' in French.

She blushed and smiled shyly and then, as before, became serious and hurried back into the inn. I watched as she left us, completely puzzled as to what was going on.

As I turned back I saw that the old woman had approached the table and was talking to me in French with a funny little smile on her face. She waved her hand toward the inn.

Unable to understand her, I turned to Noel, who was looking at her with a strange expression on his face.

Very firmly, he said "Non, madame, merci, mais non!"

The old lady spoke again, this time with an ingratiating whine in her voice, wringing her hands together. It almost seemed as if she was begging.

Once again Noel spoke firmly, "Non Madame!"

The old lady turned on her heel and stomped off toward the inn, obviously very disappointed, scowling at me with one last glance.

"What the hell was that all about?" I asked Noel.

I had never seen Noel look so taken aback.

"She wanted to know if you wanted to buy her granddaughter for an hour!"

"You're kidding!" I said. I didn't know what else to say.

"No I'm not kidding," he said. "She was offering the girl to you."

I looked at Noel, aghast. I had never been in such an awful situation before. To think the old woman had offered to sell the girl to me for an hour shocked me deeply. I guess I was pretty unworldly, growing up in Bermuda. Of course, I had heard about prostitutes and pimps when gathered together with a group of guys, but I always knew that they were just talking , not from experience but from what they had heard from others.

We hurriedly finished our food and drinks, paid the bill and left on the scooter. The day had started with the sun shining through the trees and the breeze rustling the leaves. Now, there was a sour taste in our mouths and France had been spoiled for us. We spent quite a bit of time trying to work out what had happened at the inn and why.

Was the girl really her granddaughter? Perhaps she was a war orphan that the old lady had taken in, and after the war, things were so bad that she was desperate enough to resort to selling the child. She had obviously told the girl to change into something more attractive, to make her more appealing. Whatever the reasons, it was a very sad and sordid commentary on how desperate people can become.

Every time I think of that encounter, I become very sad and wonder what happened to the young woman.

PARIS

Paris was Paris. We arrived around lunch time and parked the bike and walk around the Champs-Élysées and looked at the Arc de Triomphe. We rode around the Eiffel Tower, risking our lives in the traffic, and walked along the Seine to see the Notre-Dame cathedral. It just didn't seem the same after our cafe experience, and we definitely thought Lyon was prettier, but perhaps we were just not in the mood.

We made the mistake of sitting at one of the tables outside a smart cafe and ordering a coffee and croissant and almost fell off our chairs when the bill came. It was three times what it would have been in Lyon.

Tired and depressed we checked in at the youth hostel and at least they were only charging the usual modest fee for the night and breakfast, thank goodness. We ate at a nearby cafe suggested by the hostel manager and decided that we had seen enough of Paris and France and that we would head for Calais and the ferry to England the next day.

CALAIS-DOVER

We left Paris early the next morning knowing it would take us a good four to five hours to get to the ferry port at Calais, where we hoped to get a ferry across to Dover without having to spend the night in Calais.

The queue was quite long but we parked the Vespa and got in line, having seen the schedule that told us there was a ferry to Dover at 2.00 pm. Neither of us had seen the channel ferry operation and it was quite an eye-opener. Behind the ticket office we could see the ferry waiting at the pier and.to me, it looked like it was almost the same size as the Queen of Bermuda which used to come into Bermuda every week from New York. Ships coming into Hamilton harbour couldn't be too large or they wouldn't be able to fit through Two-Rock Passage, the narrow entrance to the harbour.

We bought our tickets and rode the Vespa onto the ferry, being directed into the large area that looked like an aircraft hangar for cars. We were told to park the bike in the area just for bikes, locked it and rushed up to the observation deck so we could watch the departure from Calais. We were like schoolchildren on an outing.

The crossing to Dover was supposed to take a little over an hour and we spent the time wandering around the ship, but mostly we just stood at the rail and marvelled at the dozens of ships that we could see travelling back and forth along the English Channel. The Channel had been described as the busiest stretch of water in the world and we could see why. At times it looked as if we must certainly collide with a ship that seemed to be headed directly for us, but our captain would adjust his speed and course and we would slide sedately behind the other ship quite safely.

Shortly after we set off we realised that the hazy white line on the horizon ahead of us was the white cliffs of Dover which caused lots of excited anticipation among the Brits on board. We were excited too, having read all about them in books and seen them in movies about the war. I had been born in Kent in the Royal Navy maternity ward in Chatham but had left with my mother in 1944 at the age of four after my father was torpedoed so had never seen the cliffs even though I was from Kent.

The cliffs are very impressive as you get closer to them, and by the time we entered Dover harbour, we felt dwarfed by them as they towered above us.

We became so interested in watching the procedure of docking the ship in Dover, watching the crew throwing lines ashore and then slowly winching the ship in, that it suddenly dawned on us that we were the only ones on deck! We hurried down to the car deck to find people already in their cars with the engines running, the loading doors open and the ramp slowly lowering. In a few short minutes we were in the line of cars and bikes leaving the ferry and joining the queue at customs. We were directed to a line for bikes only and when we got to the young customs officer we thought he would just look at our British passports and wave us through. However, this was not to be the case.

"Good afternoon gentlemen, where have you come from?"

I guess we wanted to boast a bit so we replied, both of us taking turns...

"We've come from Paris today."

"But we've actually come from Bermuda through Spain and Italy."

He appeared to become very interested.

"So you have come from Bermuda via Spain, Italy and France?"

"Yes that's right!" we both agreed, a little proudly.

"Hmm, alright then, but where did the scooter come from?"

"Oh, we bought it in Milan."

"I see. So you are really wanting to import this scooter into the United Kingdom, is that correct?"

Noel and I exchanged looks. This was not going as smoothly as we expected.

"Erm, yes, I suppose we are."

"I see," said the customs officer, not much older than us and appearing to relish his game of cat-caught-the-mouse he was playing with us.

"Well," he continued, "given the age and condition of your bike, you will have to pay an import tariff on it of £45."

"What?" we both sputtered at once.

"Yes, I'm afraid so, gentlemen." He smiled as he closed the trap.

"But we only paid £25 for it!" Noel protested.

"Sorry, gents, but those are the rules."

By this time a second customs officer had joined our line, probably in response to some hidden signal from our chap, and we had somehow been pulled off to one side. This situation was obviously something that happened all the time and they were old hands at coping with it. We were now separate from the line of bikes which flowed smoothly past us.

"Is there a problem?" our chap asked.

The two of us started to talk to each other for his benefit.

"Do we want to pay that much for it?"

"I'm not sure, I mean, would we use it very much in England?"

"What else can we do?"

At this point, the customs chap joined the conversation, and it was obvious that he had done this before.

"Look chaps, I can see you aren't trying to pull fast one over on us, and you weren't expecting this complication, so if it will help you out, I am prepared to pay you what you paid for the bike in Milan, and as the new owner I can sort out the paperwork at the end of my shift."

He smiled as if he was doing us a great favour

We crossed the English Channel with a Vespa scooter and walked out onto British soil with our packs on our backs and no scooter! Fortunately getting a lift to London was very easy because there were hundreds of cars and trucks coming off the ferries and headed that way.

As we sat in the back seat of a rather nice Mercedes estate car that was piled high with duty-free booze in the back, we quietly discussed what had just happened.

"He's done that before, you know."

"Yup, I reckon so, got a nice little business going on the side."

"Yeah."

"Oh well, we got our money back and that scooter took us over the Alps."

"It was a good bike for the price."

The lady in the front of the car next to her husband turned around to say, "We are stopping at a nice place we know if you would like to join us for an afternoon tea. Our treat."

Perhaps it wasn't such a bad day after all.

ENGLAND

We went to the youth hostel in Earl's Court, an interesting area with lots of nightlife. The hostel was in a large old building and it had lots of small rooms which accommodated two persons at a time. This meant we could stay in a room together.

We stayed there for about a week, doing the usual tourist things but we were preoccupied with other plans. Noel had plans to visit his aunt who lived in London and he wanted to stay with her for a while while he looked into joining some sort of writing school. He was very interested in a career in journalism or some other form of writing.

Similarly, I wanted to go to see my granny. My mother's mother, Bessie, had moved back to England and was living in Blackburn, Lancashire near her sister who lived in Preston.

When my mother and I had left England after my father was torpedoed in the war we had stayed with her parents in a place called Bradley Forest in Virginia. I had had a great few years there before moving to Bermuda. Living in a forest in Virginia was fun, and I soon made friends with the neighbouring boys and we ran through the forest startling deer and rabbits and dodging snakes. I vividly remember that I must have forgotten to dodge a bush called poison ivy because for about a week I couldn't go to the school in Manassas, Virginia because I was covered from head to toe in an incredibly itchy and painful rash. We used to play in a muddy creek called The Bull Run which I later discovered was a feature in a big battle in the American Civil War.

After staying at the London youth hostel Noel and I parted ways as he went to his aunt's house to stay and I headed north to Lancashire to find my granny's house. We agreed we would get back together in a month and plan the next segment of our trip around Europe. We were even talking about just going as long as we could, maybe even around the world! We set a date that we would meet back at the London hostel and set off on our separate ways.

BLACKBURN

It was the beginning of June 1960 and the weather was very pleasant. I had not seen my granny for quite a while. She was my mother's mother, and we had stayed with her and my grandfather in Virginia when we first left England. By that time they were both retired, although I do not think my granny ever worked after having her four children. Grandfather John was a very successful businessman. After he died my granny had decided to move back to England and the area from whence she came. She was an Aspinall and they had come from Preston in Lancashire.

The trip from London to Blackburn was about two hundred miles so I set off in the morning, following the directions of the hostel manager and taking a bus to the outskirts of London to where I hoped to pick up a ride from a long-distance lorry driver or a salesman.

Hitch-hiking in England was easy back then, with none of the risks of muggings and rapes that you hear about today. I soon had a lift from a lorry driver who dropped me near Birmingham where he was delivering his goods, and soon I had another ride with a salesman who was going to Liverpool. He never stopped talking for the whole trip, asking me all sorts of questions about Bermuda and seemed to think that I and everybody in Bermuda, was a millionaire.

After half an hour of walking along the road, which was not a motorway in those days, I was picked up by a guy called Gary who was a salesman and delivery driver for a large pet food supplier. He was going home and when I told him where I was going said that my granny's address was only a hundred yards from his house and he would drop me at her door.

When Gary dropped me off he gave me his address, telephone number and directions how to get to his house from my granny's house, saying that we should get together, and drove off. I looked around. There were a few modest houses on the street, some old and some new. Gary had told me that the area had once been on the outskirts of Blackburn but had been swallowed up by the town as it expanded. My granny's house looked two hundred years old. It had a rough stone exterior, very solidly built, simple with a front door and a window on each side of the door. A wisp of smoke drifted from the chimney.

I knocked on the door.

"Hello grandson!" my granny said with a smile as she opened the door. "Your mother wrote and said you would be showing up on my doorstep sooner or later. Welcome!"

Bessie Turner was 70 years old by now with grey hair and a slightly wrinkled face and couldn't have been more than five feet tall, but was from old pioneer stock. She had left England for America to start a new life with her husband John Turner. I knew that to be fooled by her size would be a mistake. When staying with them in the woods in Virginia I had watched as she walked out the door one Saturday saying, "Chicken tonight," and caught a nice fat hen that was feeding and clucking around in the back garden. She seized it and expertly wrung its neck in a split second and then proceeded to pluck it clean and gut and cook it. I never forgot that.

Inside the small cottage was basic accommodation of living room, bedroom, kitchen and small bathroom. I wondered where I was going to stay, but she showed me a set of steep narrow stairs that disappeared up into the roof of the cottage. Up there she had it fitted out as a clean and comfortable guest room. I got the impression that it had been done recently for my visit. After I settled in we sat and chatted about how things were with my mother, her daughter, in Bermuda, and then had a nice home-made stew.

The next day granny showed me the grocery shop and red phone booth on the corner at the bottom of the road where her cottage stood, letting me know in her own quiet way that I was expected to share in the cost of food. There was no phone in the cottage and if I wanted to contact anybody such as my newfound friend Gary, I would have to make sure I had change for the phone booth.

We sat and caught up on everything, and she surprised me by telling me that she was going to see her sister May in Preston the next day. She had just been waiting for me to arrive. She was seventy years old and she was going to jump onto a bus to Preston, ten miles away and stay there for a few days. I thought she was amazing. She had it all worked out. Nothing fazed her. I wouldn't have been surprised if she had told me she was going to hitch-hike to London. That night we sat in front of the glowing coal fire and chatted.

"Did your mother ever tell you about our boat on the Hudson?" asked granny.

I perked up. A boat? On the Hudson River in New York state? This sounded interesting.

"No, she never mentioned it," I replied and she proceeded to tell me the story.

My grandfather was a successful businessman, and after he and my granny had been married in 1909 and they started to have children they had bought a house situated on the Hudson River in upstate New York. They had acquired a nice motor cruiser that they kept moored in front of the house on the river and had it for a few years, but in the late 1920s they had decided to sell it and advertised it in the local papers.

The 1920s was what is known as the 'prohibition days' in America when it was illegal to make or sell alcohol. This led to all sorts of criminal activity as organised gangs tried to avoid the law and smuggle alcohol all over the United States. Many movies have been made of the notorious gangsters and their feud with the law enforcement officers, and the secret 'speak-easy' nightclubs during the 'roaring twenties'. My grandparents did not know that they were about to become a part of this culture. One day, my granny told me, a nice young man knocked on their door and asked about the boat. After a short conversation and a quick look at the boat, he handed over the required sum of money in cash and left, saying that two men would be along the next day take possession of the boat.

All went well and the boat was collected and granny and grandpa thought nothing more about the boat until about a month later, granny was looking through the New York Times and suddenly said to her husband, "John, our boat is on the front page of the Times!"

They both read the article, which included a lovely big picture of their old boat, and were shocked to find that the nice young man that had bought the boat had his picture in the Times as well, and was identified as a member of a well-known criminal gang of smugglers. The boat had been seized by a customs task force as it tried to smuggle a large quantity of illegal booze ashore from a larger boat lying offshore near New York City.

"That explains how they could afford to pay for it in cash!" said my grandfather.

They had many a good laugh and entertained their friends with the story of their rum-running boat.

With my granny in Preston, I had plenty of spare time on my hands before I had to travel back to London to meet Noel. I had gone to the bus stop with her and waved her off and then returned to the little cottage.

I didn't know anything about Blackburn but I found some books in the cottage and learned that it had been famous for a certain type of checkered cloth many years ago. It was what is known as a mill town, one of many in Great Britain in the 1600s. It also had a cathedral, said my guide book, but after seeing the cathedrals of Madrid, Barcelona and Milan I thought I would give it a miss.

According to one of the guide books, it had a lovely park. It was a nice day so I decided to walk to the corner, call my new friend Gary from the phone booth and if he didn't answer I would walk down to the centre of town and check out the park. There was no reply when I called him. In those days there was no call-forwarding or message recordings on telephones so I just decided to call him later in the evening and wandered down to the park. It was a very nice park with a pond and the usual swans and ducks, very relaxing and enjoyable so I walked around it for a while and then went back to the cottage. I stopped at the phone booth and Gary answered the phone, telling me he had just come in from a delivery run.

Gary invited me to accompany him on his run the next day and we spent all day driving around Lancashire delivering bird feed to little pet shops. He was very enthusiastic and knowledgeable about his job and he talked about bird feed non-stop for most of the time. Although I wasn't all that interested I soon became expert on what birds like mealy worms or sunflower seed and I learned that you could even buy seed cakes for certain birds, stuck together with suet or fat which was especially good to put out for birds in the winter months. It was quite good fun going out with Gary and I got to see lots of small towns all around Lancashire as we delivered feed for our feathered friends.

After three weeks I still had a week before meeting Noel at the prearranged time in London. I thought it might be a good idea to call him and, after a few attempts, I finally managed to get through to his aunt's house on the phone. When I asked to speak to Noel she told me, "Oh, Noel's gone back to Bermuda, dear."

"What?" I said, rather stupidly. I had heard her perfectly.

"Yes, dear. He asked me to let you know when you called. It was all rather sudden. He got a telephone call from Bermuda and he caught the first flight available."

"Really? Do you know what happened? Is one of his parents sick or something?" I asked.

"No dear, nothing like that. He told me to tell you not to worry, it was just something he had to sort out and he would explain everything when he saw you back in Bermuda and that he was sorry to leave you like this, but he had no way of getting in touch with you," she explained.

I relaxed slightly. At least everything was OK as far as Noel and his family were concerned, but I still wondered what on earth could have made him rush back home. I decided to put it to the back of my mind and instead of worrying about Noel, I had to decide what I was going to do.

I could keep travelling around Europe for the rest of the summer, although it wouldn't be the same on my own. I decided to take my time and figure out what to do.

I went out with Gary the next day and as I sat in the cab with him I realised that although he was a really nice guy and liked having me along as company, I really couldn't stand any more non-stop conversation about the endless variations of bird feeds that he was carrying in his van and the next time he invited me along I had to think up an excuse for not going.

"I can't go, sorry," I told him.

Obviously disappointed, he asked why not.

"I'm going to apply for a job as a bus conductor," I blurted out.

I had seen the advertisement in the local paper that evening and had thought about applying. Now I was committed. I walked down to the Ribble Valley Transport office in the town and entered the manager's office confident that I would never get the job but I would have a clean conscience because I had not lied to Gary. I had estimated that they would never hire a person who had never seen Blackburn before a few weeks ago.

"Are you good with figures? Can you do simple sums in your head, like five shillings less two and ninepence?" the manager asked me.

"Oh sure, that's easy, two and thrupence," I said proudly. I had always been good at figures and knew it.

"Well, you seem to be quite presentable and respectable looking. I like to think I am a good judge of character and I am willing to take a chance on you. You can learn the route from your more experienced conductor who will be assigned to you for the first week and then you'll be on your own," he said.

I stared at him. "You mean I've got the job?" I asked, incredulous.

"Yes, go to uniform stores, Miss Wright will show you the way, and you can start tomorrow morning on the 5.00 am run."

In a daze, I followed the efficient Miss Wright, who happened to be about fifty years old despite her title. Half an hour later I had a paper bag full of uniform, a little badge that said "McGill-Trainee", my staff instructions and a map showing how I could get from the cottage to the bus depot and garage to start the early 5.00 am run. I calculated it meant that I had to leave the cottage by 4.30 am and walk very briskly to the depot to be on the first bus of the day. Listening to Gary lecture me on bird feed all day seemed like a very attractive option at that moment.

The Ribble Valley bus company is named after the Ribble River and the valley it runs through in the Blackburn and Preston area. They had both single deck and double-decker buses. I got up at 3.30 the next morning and predictably, it was drizzling rain. There was no bus for me to catch to work because, guess what?, I was going to be working on one of the first buses of the day! I trudged miserably along to the bus depot clad in my new uniform, complete with a raincoat thank goodness and sought out instructor, Bill, who seemed to be quite a nice older guy.

"Right, John, this will be our bus from now until noon today," he announced cheerfully as we walked along the line of buses in the depot and stopped in front of one.

My heart sank. It was a double-decker.

I spent two weeks on the buses and then had to give it up. I threw in the towel, as they say. I had started the job just when the weather took a turn for the worse. I had a cold, I walked to work in darkness every morning, most times it was raining. I spent the whole shift running up and down the stairs on the bus because Bill's legs "aren't as good as they used to be". Apart from trying to remember the fares around the town I had to learn the local Lancashire dialect, although I am sure that the rough-and-ready workers who were on my bus soon learned there was a new boy on their run and he was a 'foreigner', so they used every local slang word they could think of to have fun with me.

If I was told once, I was told a hundred times, "Aye it's mizzling an' yer lookin' fair witchett the day lad," which meant, I was told by Bill, that it was hard, steady rain and I was looking decidedly drenched. There would be loud laughter all around the bus whenever I looked blankly at my jokester passenger.

After two weeks I gave in my notice and, because I was still under instruction they let me go without making me work out the usual two weeks' compulsory notice. My legs ached, my back ached, my nose was running and my pride was hurt. I had tried my best but I gave up, my tail between my legs.

My granny had gone to visit her sister again and was due back any day and I had to make a decision. I decided to head back to the London hostel and look around that city a bit more and then go across and head for Germany or Switzerland. I was a free spirit and could go anywhere I desired, as long as I could get the odd job once in a while to earn some money.

I left for London the next day.

EUROPE

I got a ride almost as soon as I got out to the main road just out of the town limits of Blackburn. The guy that picked me up was a salesman whose name escapes me now, but he was going all the way to Coventry which was a good distance so I accepted his offer of a lift when he stopped.

We got to Coventry around noontime and I decided to look around before finding the youth hostel. Coventry was famous for two things. About 800 years ago the Lord in charge of Coventry was going to make all the citizens pay a very high tax, but his wife, Lady Godiva, pleaded with him not to do this. He was not a very nice man and he sadistically told her he would not tax the people if she rode through the city naked. She cared so much for the poor people of the city that she agreed to do it, and when the people of Coventry heard what she was going to do for them, they all agreed among themselves that nobody would look out of their houses when she rode by that day. According to legend, only one man, named Tom, looked out of his window and he was forever known as Peeping Tom.

Her husband, even though a thoroughly rotten type, honoured his promise and did not raise the taxes.

Wikipedia
Statue of Lady Godiva in Coventry.

The second thing that Coventry was famous for was that it was the most bombed city in England in the Second World War. For some reason, Hitler, nice chap that he was, decided to make an example of Coventry and bombed it so enthusiastically that he destroyed the city centre, including the cathedral. After I had spent enough time admiring the statue of the lovely Lady Godiva, I wandered along to have a look at the cathedral and it was very moving, a bit like The Last Supper because although the German bombers had destroyed most of the main church, the tower and spire were still standing.

They were almost finished building a new cathedral when I was there, and it had been built attached to the ruins of the old one, almost like a gesture of defiance. It was quite a contrast because the old building had been cleaned up but left with just a few walls and empty windows standing with the steeple and right next to it was this very modern building that somehow worked. It made me feel proud to be British. Up you, Hitler!

After spending the night at the hostel I got on the road early and was picked up by a lorry driver who was going all the way to London. As soon as we got going he slowed down because he wanted to look at a lorry crash he had heard about. The road was a dual-carriageway and we looked across the centre divider and could see this flat-bed lorry on the verge on the other side of the road. It had been carrying a large load of reinforced iron bars. It was awful to look at because the driver must have had to apply his brakes hard for some reason and the whole load had slid forward and had shot right through the driver's cab, demolishing it and decapitating the driver. I could see that it really upset my driver, I guess he could sort of relate to it, being a heavy goods lorry driver himself. It was an image I have never forgotten.

After spending a couple of nights in the London youth hostel at Earl's Court I decided to go a-wandering, as they say. I didn't want to go back to France so I looked at the ferry timetables and saw that I could go from Dover to Ostend in Belgium, and from there I could decide on either Holland, Denmark or Germany.

I got a lift to Dover very easily, there were always lots of people headed in that direction. I arrived in Dover in the afternoon and then discovered that there was an evening ferry to Ostend but instead of the short trip of about ninety minutes to Calais, the crossing to Ostend was over 4 hours. I was worried about arriving in Ostend late at night but decided I would give it a go.

I got lucky. On the Ostend ferry, I got to talking to an American chap who was in the American army and stationed in Germany in the city of Frankfurt. I do not recollect his name but he told me he was driving to Frankfurt direct from Ostend and he could almost guarantee me a job at the army establishment where he worked because I had my U.S. passport with me. I jumped at the chance of a ride to Frankfurt, a place of which I had no knowledge but it was a start. We arrived in the early hours of the next day, which was now the first week in July.

I'm not sure what I expected to find in Frankfurt but I was disappointed. I think a lot had to do with the fact that I was now travelling alone and it wasn't the same as being with my buddy Noel. Frankfurt seemed to be a dull and uninteresting city and I had a hard time being enthusiastic about exploring it.

The US Army officer who had given me a ride had dropped me off at a hostel and it was quite nice and reasonably cheap. It was early in the morning so I had time to kill before checking in, so I just wandered around a little and had lunch in a cafe. I did not want to apply for a job at his place of work before I had a chance to clean up and get a good night's sleep.

The next morning I went to the I.G.Farben building where the US Army had set up offices after the war. I found it interesting that they were still there fifteen years later, and the US Army had their headquarters there. It was a huge complex of buildings and it took me a while to get directions to the right entrance and then find the right person. I found an officer who knew the guy who had given me the ride from Ostend and finally I found myself being taken on as a general helper in the Officer's club.

I was pretty low on money and decided to stick it out for a while. I worked for the manager of the club and went in every morning to be assigned menial tasks including cleaning up the mess that had been left from the night before. It wasn't too bad, I simply used a vacuum cleaner on the carpets, straightened up chairs and tables and things like that. There were quite a few of us and the Americans paid very well for such light duties. I guess you can be generous when you have won the war.

The I.G.Farben company was a big industrial conglomerate that had produced many things for Hitler's war effort and I remember thinking that it was a bit strange for me to be working in a building that might have contributed to my father being torpedoed.

After two weeks of this mind-numbing work, I decided I was going to quit and gave my notice to work two more weeks and then go back to England and then Bermuda. I had been away for four months and was homesick. I still had my ticket home from London that I had brought with me, so I went back to London at the beginning of August and flew home on BOAC.

BERMUDA

Back in Bermuda, I moved in with my mother at the large family house on Jew's Bay. The day after I arrived I went to Hamilton to say 'Hello' to all my old workmates at American International, and then I walked into the office of my old boss Mr Buswell. He asked me to sit down and tell him a bit about my travels and then he said, "Well, it's nice to see you back John, do you have any plans for more travel in the near future?"

"No sir, I have had enough travelling for a while."

"Good, I'm glad to hear it. Would you like your old job back?"

Just like that, I had my old job back. It gave me a good feeling for him to offer me a job so quickly. They had obviously valued me. I started the next day.

I contacted Noel the day I had arrived but he was reluctant to talk on the phone so I had to wait until the weekend to go to his house to see him. It was great to see him again, but his girlfriend Florence was there and he didn't want to say much in front of her, but we went outside and he told me that they had decided to get married and that was why he had come back. I got the impression he wasn't telling me everything but I left it at that. I think it involved Florence and I did not want to push him on the subject.

We kept in touch but we never got up to any more exciting things because he was always busy with Florence, and I became involved with new friends at my job and we didn't see each other as much as we used to. Once in a while we would go down to the shed on the dock and sit with a beer and talk about the old days and our adventures in Europe.

Christmas came and went and life continued in a routine sort of way, then one day I got a very brief phone call from Noel saying that he had decided to go to London to study journalism and he didn't know when he would be back. It was all very strange and mysterious, especially because he was leaving on the BOAC flight that very evening.

I kept in touch with Noel's parents, and they would give me updates as to what he was doing and I got the odd letter from him once in a while. He had signed up at a London college for journalism and I was happy for him because while we were travelling he always spoke of wanting to write one day.

THE END

Several months went by. It was early 1961, and I was sitting at my desk at American International on Pitts Bay Road, Bermuda. I had my head down, working on one of the spreadsheets I liked to play with. The object was to convert the dozen or so foreign currency amounts of total reinsurance premium figures from the different affiliates that our company dealt with, convert them all to US dollars and then achieve the miracle of making all the column totals when added across, agree with the grand total. It always made me feel good when it worked out.

I sensed a presence and looked up. It was Gordy, standing in front of my desk. I remember thinking, "What the hell is Gordy doing here? He never comes here to my place of work."

As I looked up, he stood there, looking at me nervously.

"Hey, Gordy, what's up?"

"Hey, mate. I just wondered if you had heard about Noel?"

"What do you mean?"

"There's been an accident."

Four words that freeze your heart.

"An accident?" I asked, not wanting to hear the reply that I dreaded.

"Yeah, mate, I knew you would want to know, you being his mate and all that."

"What happened Gordy?"

"He and another guy were riding a scooter. I think they were in Italy or going to Florence in Italy. A big car transporter pulled out in front of them and they hit it."

"Oh shit. Is he all right?"

"No, mate. They both died."

I stumbled to the Gents toilet, went into the cubicle, closed the door and wept.

Printed in Great Britain
by Amazon